RESEARCH ETHICS IN THE LIFE SCIENCES

RESEARCH ETHICS IN THE LIFE SCIENCES

Michael J Kuhar, PhD

© 2018 Michael J Kuhar, PhD
All rights reserved.

ISBN: 1977695175
ISBN 13: 9781977695178
Copyright attribution: <a href='https://www.123rf.com/profile_boygointer'>boygointer / 123RF Stock Photo</a>

Dedication

This work is dedicated to the scientists, ethicists, administrators, and others who have contributed to and developed the field of research ethics.

Acknowledgments

Many colleagues provided helpful discussions and comments. They include Lou Ann Brown, Mary Delong, Michael Deryck, Kathy Griendling, Randy Hall, Kathy Kinlaw, Denyse Levesque, Eddie Morgan, Elizabeth Strobert, Roy Sutliff, Larry Young and Hal Wiedeman. Stuart Lustman helped edit the manuscript. Rita McMaken assisted with cover and text design and organization.

The author has been supported over many years by the Yerkes National Primate Research Center of Emory University, the Emory University School of Medicine, the Georgia Research Alliance, and the National Institutes of Health.

Contents

Introduction · xiii

1 Mentor-Trainee Relationships · 1
   The mentor - trainee relationship · · · · · · · · · · · · · · · · · 1
   Obligations of mentors · 2
   Obligations of trainees · 4
   Problems for mentors · 5
   Problems for trainees · 6
   Cases · 7
   Summary · 9
   References ·10

2 Premises, Experimental Rigor and Reproducibility · · · · · · · ·12
   Reproducibility ·12
   The premise of the research must be
   critically examined ·13
   Applying rigor ·14
   Transparency ·14
   Relevant biological variables · · · · · · · · · · · · · · · · · · ·14
   Authentication of biological and chemical resources · · ·15
   Training ·15

	Cases · 15
	Summary · 17
	References · 19

3   Animal Experimentation · 21
    Regulatory agencies and guides · · · · · · · · · · · · · · · · · 21
    Some background principles · · · · · · · · · · · · · · · · · · · 23
    IACUC applications · 25
    Violations · 26
    Cases · 27
    Summary · 29
    References · 30

4   Human Experimentation · 31
    Background of the protection of human subjects · · · · · 31
    Requirement for training · 33
    Scope of human experimentation · · · · · · · · · · · · · · · · 34
    The IRB and its reviews · 34
    The protocol · 36
    Continuing Responsibilities · 39
    ClinicalTrials.gov · 39
    Cases · 40
    Summary · 42
    References · 43

5   Data Acquisition, Management, Sharing and Ownership · · 45
    The notebook · 45
    What goes into the notebook? · · · · · · · · · · · · · · · · · · 48
    Who owns the data and who is responsible for it? · · · · 49
    The Bayh-Dole act for patent issues · · · · · · · · · · · · · · 50
    Confidentiality · 51
    Sharing data · 51

		Big data	52
		Cases	52
		Summary	54
		References	56
6	Authorship		58
		What constitutes authorship?	58
		Multiple authors and positions in the list	60
		Corresponding author	60
		Dealing with disputes	61
		Misconduct	61
		Conflicts of interest	61
		Cases	61
		Summary	63
		References	64
7	Peer Review		65
		Purpose of peer review	65
		Requirements for sound peer review	66
		Problems with peer review	68
		Cases	70
		Summary	71
		References	72
8	Research Misconduct		73
		Definitions of misconduct in research	73
		Identifying and investigating misconduct	74
		Causes of misconduct	76
		Protecting whistleblowers	76
		Are you considering reporting misconduct?	77
		How can researchers help avoid problems?	79
		Cases	80

	Summary · 83
	References. · 84

9   Conflicts of Interest and Commitment · · · · · · · · · · · · · · · 86
       Financial conflicts of interest · 87
       COIs in reviewing grants and contracts · · · · · · · · · · · · 88
       COIs in authorship · 89
       COIs in reviewing papers submitted for publication · · 89
       Conflicts of commitment · 89
       Cases · 90
       Summary · 91
       References. · 92

10   Ethical Behavior Among Colleagues · · · · · · · · · · · · · · · · · 93
       Time and effort for growth · 93
       Give credit where it is due · 94
       Being a good collaborator · 94
       Diversity · 95
       Use positive and supportive language · · · · · · · · · · · · · 96
       Practice the golden and platinum rules · · · · · · · · · · · · 96
       Conflicts among colleagues · 97
       Avoid bruising colleagues · 97
       Human nature: both positive and negative influences · 98
       Take some time before responding. · · · · · · · · · · · · · · 100
       Unethical colleagues · 101
       Can we learn to be better colleagues? · · · · · · · · · · · · 102
       Cases · 102
       Summary · 105
       References · 107

Notes · 109
About the Author · 113

Introduction

This volume is best described as a short guide. It is mainly practical and covers the basics of ethical behavior in a research setting. It is primarily for those recently entering the research arena such as advanced undergraduate students, medical or predoctoral students, postdoctoral fellows and others, although it can be a good reference work for the more experienced as well. It may also be valuable at smaller institutions that do not have other more developed options.

It is not a detailed literature review. Rather, it is more of a focus on what researchers need to do (1-3). Only representative references are given, but nonetheless, these readings provide much additional information.

This book presents guidelines and topics that are fairly well known as well as some that are newer. Newer guidelines are those that relate to the premises underlying research, and the reproducibility of data by improving rigor and transparency. Also, a chapter on co-worker/collegial ethics is included.

Michael J Kuhar, PhD

Throughout this book, one of the recurring themes is the need for training. Allotting time for ethical training is, in practical terms, required in a research career. Because regulations are sometimes improved or added, continuing training will be required.

While this guide will highlight important ethical issues, it cannot cover all practical situations. The variety of possible problems, projects, techniques, types of data, funding rules, and institutional rules are great. Researchers will need to discuss, with their supervisors, institutions and funding sources, the various required procedures and training. This guide will hopefully help in that process. Medical ethicists are an invaluable resource in many situations, especially in human experimentation or when confronted with ethical dilemmas. Some ethics courses are available online as well (4). In situations where procedures are not completely specified by laws or regulations or strict policies, this text attempts to recommend actions that that will be most helpful to investigators. They will hopefully prevent future problems.

Note that the terms trainee and mentee are used interchangeably. Also, cases are used to illustrate a variety of principles (5-9). A study of cases is a well-known and effective learning tool. None of the presented cases refer to a specific event or person except for a very few that have been widely publicized or quoted from other sources.

The author welcomes comments.

Some Readings

1. Emory's Guidelines for Responsible Conduct of Scholarship and Research. http://policies.emory.edu/7.9, accessed on March 8, 2017.
2. Guidelines for the Conduct of Research in the Intramural Research Program at NIH. https://oir.nih.gov/sites/default/files/uploads/sourcebook/documents/ethical_conduct/guidelines-conduct_research.pdf, accessed on March 8, 2017.
3. https://oir.nih.gov/sourcebook/ethical-conduct/research-ethics, accessed on March 24, 2017.
4. An example: https://about.citiprogram.org/en/series/responsible-conduct-of-research-rcr/, accessed on July 20, 2017.
5. Annual Review of Ethics Cases. https://oir.nih.gov/sourcebook/ethical-conduct/responsible-conduct-research-training/annual-review-ethics-case-studies, accessed on March 8, 2017.
6. https://www.niehs.nih.gov/research/resources/bioethics/whatis/index.cfm, accessed on March 24, 2017.
7. NSF on Research Ethics. https://www.nsf.gov/bfa/dias/policy/rcr.jsp, accessed on March 24, 2017.

8. http://ethics.iit.edu/eelibrary/case-study-collection, accessed on April 7, 2017.
9. http://research-ethics.net/discussion-tools/cases/, accessed on April 7. 2017.

1

Mentor-Trainee Relationships

THE MENTOR - TRAINEE RELATIONSHIP

Mentors are colleagues who are successful, mature, and ready to impart skill and vision to trainees (Note that the terms mentee and trainee are used interchangeably). This mentor-mentee relationship is one where the mentors guide and instruct the trainees in how to get to the next level in their career. In academia, it is a mutually beneficial relationship where the mentor benefits from the efforts of the trainee, and where the trainee is taught the related processes of grant writing, research, and publication of the results. Each of these processes is complex with many skills to master. Mentoring trainees is a serious undertaking that can be time consuming for both mentor and trainee.

Trainees want to move to the next level in their career. They are prepared to work diligently, carry out the required tasks, and take direction from a mentor.

Mentoring can occur at many levels and is usually time sensitive. There is the dissertation advisor and the PhD student, the faculty member and the post-doc, the senior scientist and the young investigator, and yet others. These are different challenges for mentors because the trainees vary in their background and skills. Usually there is one primary mentor, but the trainee may gain significant experience from others such as coworkers. A good idea is to have informal mentors, those with whom the trainee can chat and get advice.

Before the relationship begins, there should be a clear discussion that leads to a mutual understanding of the responsibilities and desires of both the mentor and trainee.

Many groups have put in place various offices, policies, and goals for mentors and trainees. These include the Office of Research Integrity (1), the Intramural Research Program at the NIH (2), NIH training programs (3), and various programs at a variety of institutions (4, 5).

OBLIGATIONS OF MENTORS

Mentors learn their skills from their mentors and from other sources. Guides to mentoring can be found at several websites (6 - 9).

The following are important issues for mentors. In many cases, these duties are shared by training program staff or various institutional offices.

1. Mentoring requires setting aside regular time to interact with the trainee. This means that there is a limit to the number of trainees that one can mentor well.
2. Good communication must be developed between mentors and trainees. The trainees need regular feedback on their progress. Part of this is receiving motivation so that the trainee's tasks are correctly viewed as progress towards their goals.

3. A suitable research project must be selected. These include projects that aim to satisfy everyone's needs, are feasible and instructive to trainees.
4. Mentors are expected to stay abreast of important current topics in science such as the material in Chapter 2 and convey that to trainees.
5. An important process, carried out with the supervision of mentors, is a review of primary data, and how it is obtained, analyzed, kept, stored, and retrieved.
6. Training in the writing of scientific papers and proposals of various kinds is essential. This includes addressing ethical issues of authorship and giving credit where it is due.
7. Training in all aspects of the responsible conduct of research is expected. Some funding agencies require this.
8. Training in how the trainees can become mentors is also needed. Trainees learn by example, and mentors are role models in many ways.
9. Mentors provide guidance in other aspects of science including attending meetings, presenting findings, networking with others, assessing job opportunities, sharing knowledge of available jobs, and providing general guidance about career advancement. It has been said that science is a "contact sport" and mentors should try to provide opportunities for trainees to meet peers, potential employers or others that are needed for the next step in the development of the trainee. Also, advice on how to be a good collaborator would be beneficial.
10. Later in trainees' careers, letters of recommendation and additional guidance will often be needed in a timely manner.

OBLIGATIONS OF TRAINEES

Training is complex and different stages of training will have different requirements. Graduate students have hurdles such as qualifying exams and dissertation approvals. Post-doctoral fellows are expected to have more experience and productivity and be more ready to transition to independence. Trainees need to know where they stand in the larger scheme of things and what is required of them at that stage. All trainees share some general obligations.

1. The relationships between mentors and trainees are obviously critical. Trainees need to be sure that their goals are understood by the mentor and are acceptable, and this should be discussed before the work is started.
2. Being a passive trainee and expecting that opportunities will automatically come may result in disappointment. In general, graduate training and training beyond that requires a more independent and active approach to career development. Trainees need to be proactive in finding their way to the next step in their careers.
3. Each of the responsibilities of mentors carries a reciprocal responsibility for the trainees. If mentors are expected to teach, the mentees must be open to it and do a good job at learning. Whatever the trainees expect, they must be prepared to use it effectively when it comes.
4. Preparation is part of using the mentors' time well. For example, when planning to review primary data with the mentor, a pre-meeting examination and organization of the data will be helpful. Trainees need to be proactive with mentors.

5. Mentors realize how important communication is, and trainees vary in their communication skills; some trainees must practice, practice and practice. Writing is critical, and developing writing skills has to be done by trainees. This is often a back and forth process.
6. Trainees must be open minded enough to accept and implement criticisms and suggestions from the mentor.
7. Training is not just for the moment. Some information and skills must be carried forward. These include responsible conduct of research which covers animal and human welfare, data management, authorship, publication practices, and other skills and knowledge.
8. Trainees have to take the initiative to keep up with developments in their fields, make efforts to network, and seek out developmental opportunities as they progress. Mentors can be reassuring, encouraging, and guiding in this process.
9. Once a good mentor is found, don't let her/him go. Stay in touch because they can always be helpful in future challenges.
10. Mentors are people too. Be polite, respectful and even helpful if possible. The pressures they feel might be greater than the pressures on trainees.

PROBLEMS FOR MENTORS

Sometimes trainees don't meet expectations. In that case, the problem should be addressed with the trainee quickly. What is the problem and has it been clearly told and understood? How can it be solved? Can trainees get input from others? Are the solutions practical and reasonable? What steps will give the trainee the most help? Should the trainee be moved to another position?

Mentors also need to self-assess and ask if they have done their jobs well.

PROBLEMS FOR TRAINEES

1. A number of problems can arise for mentees such as technical problems in the laboratory, losing direction and focus, etc. There are often places to get help. When a problem arises for graduate students, there are the mentors, oversight programs, committees, and other safeguards that have been put in place for help and advice.
2. Other trainees such as postdoctoral fellows may have less protection and guidance. This has been recognized (10, 11) and some institutions have an office of Postdoctoral Education (12, 13) whose goal is to offer broad help to fellows.
3. Problems with mentors are best dealt with by talking directly with mentors. Try talking it out and work at this. If this isn't effective, then there may be other options, even changing mentors. But, depending on circumstances, changing mentors could be costly. For example, it may be difficult to find another mentor. If a trainee has spent some time in a position within a lab, others will expect them to show progress or a publication for that time. If a trainee has not invested much time, a change may be easier from that perspective. Keep in mind that changing mentors will be noticed at least early in a career and that the trainee will be asked about it in the future. There can be compelling reasons for changing mentors such as a better opportunity.

Some trainees become disgruntled or unhappy with mentors for a variety of reasons. They may feel neglected compared to other trainees. It may be said or felt, for example, that "He was right but I didn't like the way it was done." While students deserve respect in their role as trainees, mentors deserve respect as well. Everyone has a personality. Mentors will tend to deal with issues in ways they feel comfortable with, and that should be respected.

CASES

Case 1. Changing mentors is a best solution.

An example of a graduate student who was not meeting expectations follows. Dr. M has noticed that the data coming from a student was very erratic and not reproducible. Dr. M knew from his experience with previous trainees that something must be wrong because these experiments should be working and provide better data. After talking with the student, a useful fact emerged. The experiments that were failing utilized live animals and the student could not get used to working with and handling the animals. The student knew before he/she started that he/she was afraid of animals, but thought that with time the fear would be overcome. But it seemed like the opposite was happening and the student was becoming even more fearful and more incapable of injecting and handling the animals. A suggestion emerged; transfer to a lab with similar interests that did not use animals but rather used cells in culture. Another faculty member was identified, and after discussions, and a few signed documents, the student switched to the new mentor with a slightly different experimental approach. Ultimately, the student was very successful. Fortunately, because the first mentor became aware of the problem early on, very little training time was lost.

Case 2. Unexpected changes in training.

Suppose a trainee is not getting the training he/she expected in spite of the clear discussions and agreements made with the mentor previously. When the issue is brought up with the mentor, the mentor says that because of funding changes and new advances, the direction of the lab had to change, and the training needed has to evolve as well.

What are the trainee's options? How should the trainee analyze the situation?

Case 3. Polices must be fair.

A new research institute is being formed due to a very large donation from a private sponsor. The institute will be affiliated with a major university for mutually beneficial reasons. The university will be a source of talented trainees and investigators.

The new institute is developing a policy on mentor-trainee relationships. Accordingly, three committees are organized. One is made up of graduate students, another of post-doctoral fellows, and the last one of faculty.

In the policy drafts, each committee is likely to stress different parts of the mentor-mentee relationship. What do you think that each of the different committees will value most?

Case 4. Trainees can't get along.

The mentor of a team sees that some members can't get along with each other. On one or two occasions, the conflicts have been great enough that productivity was affected. Should the mentor intervene or just accept the problem as an unavoidable fact of life? If yes, how should the mentor intervene? What tools could be used (Chapter 10 may have some hints)?

SUMMARY

At the core of scientific training is the mentor-mentee relationship. Mentoring has a major impact on trainees and successful mentoring is almost always found in the careers of successful scientists. The goal of mentoring is the advancement of the trainee to the next level in their career. Mentors have many responsibilities to trainees, and similarly, mentees have many reciprocal responsibilities to the mentors. Trainees need to be proactive in their growth and professional development. Both mentors and trainees need to work out a variety of problems when they occur, such as the failing trainee, or perhaps the failing mentor. Solutions to such problems will require discussions and perhaps compromises.

REFERENCES

1. Steneck, NH https://ori.hhs.gov/sites/default/files/rcrintro.pdf, accessed on February 27, 2017.
2. Guidelines and Polices for the Conduct of Research in the intramural Program at NIH. https://oir.nih.gov/sites/default/files/uploads/sourcebook/documents/ethical_conduct/guidelines-conduct_research.pdf, accessed on February 27, 2017.
3. Training at the NIH: https://researchtraining.nih.gov/programs/training-grants, https://researchtraining.nih.gov/programs/fellowships, https://researchtraining.nih.gov/programs/career-development, and https://researchtraining.nih.gov/programs/other-training-related, accessed on October 31, 2016.
4. The Emory University link to Graduate training, an example: https://researchtraining.nih.gov/programs/other-training-related, accessed on October 31, 2016.
5. An example of a Link to Postdoctoral training at Emory University: http://medicine.emory.edu/geriatrics-gerontology/education/post-doctoral-training.html, accessed on October 31, 2016.
6. Mentoring seminar at HHMI: http://www.hhmi.org/sites/default/files/Educational%20Materials/Lab%20Management/entering_mentoring.pdf, accessed on November 7, 2016.
7. https://www.nap.edu/read/5789/chapter/2#4, accessed on November 11, 2016.
8. http://www.nature.com/nature/journal/v447/n7146/full/447791a.html, accessed on November 11, 2016.

9. http://cimerproject.org/#/, accessed on February 9, 2017 and https://nrmnet.net, accessed on August 8, 2017.
10. http://www.the-scientist.com/?articles.view/articleNo/44874/title/Opinion--The-Postdoc-Crisis/, accessed on November 7, 2016.
11. http://www.sciencemag.org/careers/2014/07/stressed-out-postdoc, accessed on November 7, 2016.
12. https://med.emory.edu/postdoc/, example of an office whose focus is postdoctoral fellows, accessed on November 7, 2016.
13. http://postdoc.hms.harvard.edu/, another example, accessed on November 7, 2016.

2

Premises, Experimental Rigor and Reproducibility

REPRODUCIBILITY

Reproducibility of data is essential for scientific discovery and progress. Publications cannot serve as foundations for future work unless they are repeatable and reliable. Moreover, and not of minor importance, a researcher's career can stall if his/her work is not something others can agree on.

It has been reported that many key papers in some fields cannot be reproduced. This has been distressing and is unacceptable. It has been attributed primarily to deficient procedures and training rather than misconduct. Some groups and funding agencies, including the NIH, have addressed the issue or reproducibility and have imposed changes in the way experiments in grants are approached and written (1-7). Performing experiments that are solid, reproducible advances, and giving funding agencies what is promised for their money, has become an ethical issue.

The NIH has decided that certain strategies that impact on reproducibility must be addressed in grant applications. This decision has been endorsed by many journal editors (5, 7, 8), and it seems submitted papers will have to address these same issues.

The attempt to improve reproducibility of research relies on several factors including critiquing the premise of the work, carrying out rigorously designed experiments, and clearly and transparently describing the experiments so that they can be reproduced. Part of this is to identify variables such as biological reagents, chemicals and other reagents. The NIH offers examples of how these issues are to be addressed in grant applications, and some online training in rigor and reproducibility can be found (4, 9). These factors are interrelated and can impact on each other.

In the past, repeating others' key work had a lower value within laboratories. No one got much for a paper that says they failed to replicate so and so. Most felt that to be competitive, researchers have to move ahead to new problems and discoveries where there is greater potential value and rewards; failures to repeat weren't considered much. But these attitudes seem to be changing. There are journals and websites where issues on repeating others' work are welcome (10). It seems that there is a culture change in this regard.

THE PREMISE OF THE RESEARCH MUST BE CRITICALLY EXAMINED

The premise is an evaluation of the previous body of scientific work that forms the foundation for the new experiments. Are the earlier experiments that drive the new experiments convincing? Have they been repeated? What are their strengths and weaknesses? Are there important factors that have not been considered in prior research? Do key parts of them need repeating? Does the investigative team agree with their interpretation?

Scientific premise is not the same as scientific significance which refers to importance. If the premise of the research is not sound, then the project is on shaky ground to begin with. The premise is the previous work that informs the experimental strategy of the new work (11). In NIH grant applications, it should be addressed in the Research Strategy Section.

APPLYING RIGOR

This means that researchers are applying the highest standards to their experimental design (12). This requires a consideration of things such as experimental logic, proper control groups, statistics, power analysis, steps to avoid bias such as randomization or blinded analysis, reliability of reagents, and more. A well designed experiment along with proper analyses is fundamental to doing good science. Inadequately designed experiments are themselves ethical problems.

TRANSPARENCY

This means that the methods are described completely enough in grants and publications so that other scientists can understand and reproduce them. A complete description includes not only detailed procedures but also details on biological variables, and biological and chemical reagents. Different fields of research may have their own standards and particular variables to deal with. References 6-8, for example, describe how some journal editors interpret this, and various scientific societies have policy statements as well.

RELEVANT BIOLOGICAL VARIABLES

A part of rigor and transparency is to be completely clear about relevant biological variables such as the animals used, their sex (13, 14), weight, age, strain, treatments, conditions of maintenance, and anything else that is needed to repeat the work (15). The importance of sex as a biological variable is widely recognized (16), and studying only one sex would require a justification.

IACUC approval should be stated. If clinical research is being carried out, researchers will have to comply with IRB regulations and this should be stated. All of this needs to be thought out, and clearly and completely written in the laboratory notebook, or in grant applications or in publications. Again, different fields of research may have different variables that need to be considered.

AUTHENTICATION OF BIOLOGICAL AND CHEMICAL RESOURCES

Sometimes investigators use reagents or tools that may degrade over time or vary from lab to lab. These could include antibodies, cell lines along with passage number, sera, tissues, hormones, etc. If this is the case, the experimenter is required to describe the status of those reagents so that the readers of their work understand their current validity and quality. Different laboratories need to develop a consensus about how to do this. It is required by many funding sources and journals that reagents are shared, when possible, with others (17).

TRAINING

Training in these topics needs to be sought out and offered. Training usually includes lecturing as well as a study of cases from the field. If a researcher is funded by an NIH training grant, then training will be required. Also, guidance from mentors and colleagues is essential and needed. NIH websites are useful as well (3, 4).

CASES

Cases 1. On rigor and transparency.

Consider the following methods section in a paper. "Sixteen rats were prepared for drug self-administration. They resided in animal care facilities at the institute. After stabilization of lever pressing for cocaine, the animals were injected (ip) with an anti-cocaine drug expected to be a useful medication for cocaine abusers. It was given in ascending doses and the data were analyzed by appropriate statistical methods."

As described, is the experimental design rigorous? Why? Are enough details given so that you can reproduce what they have done? If not, what is needed? Are there any other ethical issues?

Case 2. Is the following a rigorous and transparent description?
"Male and female mice will be randomly allocated to experimental groups at age 3 months. At this age the accumulation of CUG repeat RNA, sequestration of MBNL1, splicing defects, and myotonia are fully developed. The compound will be administered at 3 doses (25%, 50%, and 100% of the MTD) for 4 weeks, compared to vehicle-treated controls. IP administration will be used unless biodistribution studies indicate a clear preference for the IV route. A group size of n = 10 (5 males, 5 females) will provide 90% power to detect a 22% reduction of the CUG repeat RNA in quadriceps muscle by qRT-PCR (ANOVA, α set at 0.05). The treatment assignment will be blinded to investigators who participate in drug administration and endpoint analyses. This laboratory has previous experience with randomized allocation and blinded analysis using this mouse model [appropriate references]. Their results showed good reproducibility when replicated by investigators in the pharmaceutical industry [appropriate references would be added]." (18)

Case 3. Considering Premises.
You are preparing a grant on a brain chemical. Your first and last sentences are: "Chemical X is an important signaling molecule in

brain. .... This project will clarify the role of chemical X in neuroplasticity."

Between these two sentences, you must build your premise. What is the purpose of a premise? What kinds of specific issues/questions/background must be addressed in this space?

Case 4. Is the premise solid enough?

You are writing a grant application in an area related to your previous work, but it is also somewhat new. In setting up your new specific aims, you review the literature searching for support and to justify your aims. But you notice that a previous segment of the work was done solely by one lab with an antibody that they prepared. This work has not been reproduced by others and you have a concern because the antibody used was and is now very old. Also, you are concerned about the initial characterization of the antibody.

This work is an important part of your justification. What can you do to be sure that this previous work is sound, and therefore that your experiments have a solid premise?

SUMMARY

To improve the reliability of published science, several issues have been stressed and earmarked for elaboration and attention. One is to have a sound scientific premise behind the project; are the bases for the experiments sound and worthy of the time to be devoted to it? How reliable is the previous work? Does some of it need to be repeated? Regarding the new experiments, rigorous experimental designs and the best scientific approaches are required; lack of rigor is a fatal flaw. Another issue is that the experimental procedures and protocols must be written clearly enough and with enough detail so that anyone could repeat the work;

this requires adequate descriptions of procedures, analyses, and biological variables, as well as authentication of biological and chemical resources. Each of these issues must be addressed in NIH grant applications, and to implement these recommendations properly, study and training is needed. Setting aside time at laboratory meetings for discussions of these topics should be helpful.

REFERENCES

1. F. Prinz, T. Schlange, and K. Asadullah, *Believe it or not: how much can we rely on published data on potential drug targets?* Nat Rev Drug Discov, 10 (2011), 712.).
2. F. S. Collins and L. A. Tabak, *Policy: NIH plans to enhance reproducibility*, Nature, 505 (2014), 612–613.
3. National Institutes of Health Office of Extramural Research, *Rigor and reproducibility*, http://grants.nih.gov/reproducibility/index.htm, accessed on September 14, 2016.
4. National Institutes of Health Turning Discovery into Health, *Principles and guidelines for reporting preclinical research*, http://www.nih.gov/research-training/rigor-reproducibility/principles-guidelines-reporting-preclinical-research, accessed on March 1, 2016.
5. Kuhar, MJ. *Letter from the Editor-in Chief: Irreproducible results and NIH actions* Journal of Drug and Alcohol Research, 5 (2016).
6. S. C. Landis, S. G. Amara, K. Asadullah, C. P. Austin, R. Blumenstein, E. W. Bradley, et al., *A call for transparent reporting to optimize the predictive value of preclinical research*, Nature, 490 (2012), 187–191.
7. M. McNutt, *Reproducibility*, Science, 343 (2014), 229.
8. McNutt, M. Journals unite for Reproducibility. Science, 346 (2014) p679.
9. https://www.nih.gov/research-training/rigor-reproducibility/training, accessed on September 14, 2016.
10. Go forth and Replicate. Nature, Vol 536, Aug 25 2016, p 373.

11. https://nexus.od.nih.gov/all/2016/01/28/scientific-premise-in-nih-grant-applications/, accessed on September 30, 2016.
12. https://nexus.od.nih.gov/all/2016/01/28/scientific-rigor-in-nih-grant-applications/, accessed on September 30, 2016.
13. https://nexus.od.nih.gov/all/2015/12/11/what-does-it-mean-to-consider-sex-as-a-relevant-biological-variable-in-your-nih-grant-application/, accessed on October 3,2016.
14. http://orwh.od.nih.gov/research/sex-gender/ accessed on September 30, 2016.
15. https://nexus.od.nih.gov/all/2016/01/29/consideration-of-relevant-biological-variables-in-nih-grant-applications/, accessed on September 30, 2016.
16. Bale TL and Epperson CN. Sex as a Biological Variable: Who, What, When, Why and How. Neuropsychopharmacol online Pub 26 Oct 2016, 1-11.
17. https://nexus.od.nih.gov/all/2016/01/29/authentication-of-key-biological-andor-chemical-resources-in-nih-grant-applications/, accessed on September 30, 2016.
18. from http://grants.nih.gov/reproducibility/index.htm, accessed on September 4, 2016.

3

Animal Experimentation

REGULATORY AGENCIES AND GUIDES

If researchers are using animals, and particularly if they are supported by federal funds, which is the assumption in this chapter, they are required to do several things. In general, they must show that their research is worthwhile and that the use of animals can be justified. Also, they must be trained so that the animals are treated as humanely as possible. Researchers must agree to oversight as well.

The use of animals in research is regulated by the 1966 Animal Welfare Act (revised 1990) and the 1985 Health Research Extension Act. The Guide for the Care and Use of Laboratory Animals is a needed reference (1). This Guide assists both the investigator and the institution in carrying out proper procedures. Research institutions must also identify an individual responsible for the oversight of proper animal care.

There are additional federal and voluntary oversights in the use of animals in research. OLAW (Office of Laboratory Animal Welfare), the USDA (US Department of Agriculture), and AAALAC (Association for Assessment and Accreditation of Laboratory Animal Care) require that animals at research institutions receive proper care. Oversight involves site visits and sometimes corrective actions. Moreover, OLAW-approved institutions are expected to "self-regulate" which is to take responsibility for enforcing regulations. Institutions must be certified by OLAW to receive PHS funding. Be aware that regulations are sometimes updated; the USDA issued updates in June of 2017.

The responsibilities of individual researchers have been stated:

"Before conducting research involving animal subjects, researchers must develop a detailed Animal Study Proposal that is approved by an animal care and use committee (ACUC). The ACUC has responsibility for ensuring that the proposed research follows all pertinent regulations governing the ethical use of animals in research. This includes ensuring that personnel are properly qualified to conduct the study and trained in the specific procedures that are required." (2)

This last sentence speaks directly to training. Researchers should consult the Guide (1) and their respective institutional animal care and use committee (IACUC) to be sure they are qualified and appropriately prepared to conduct research involving animals. This requires not only training (even hands on training) but also refresher courses. The IACUC can suspend a project if violations are found. For additional guidance, investigators may refer to the NIH website that describes PHS policies (3) such as identifying an institutional official (IO) and describing the self-regulation of animal care. The Office of Research Integrity (ORI) has a relevant publication (4) that covers many aspects of animal use. OLAW and the Applies Research Ethics National Association (ARENA) also publish a guide for the IACUC aptly called the Institutional Animal Care and Use Committee Guidebook (5).

The compositions of IACUCs have been carefully considered. According to PHS regulations, IACUCs should have at least five members. They must include a qualified veterinarian with authority to regulate animal use at the institution, a practicing researcher who uses animals in research, a member who is not using animals and whose focus is nonscientific (for example, law or ethics), a chairperson, and a nonscientist that represents community interests. Additional members may be appointed as well (5) although there are limitations on the numbers of members from the same administrative unit. Researchers need to be aware of all of these groups, publications, and regulations (6).

Every investigator using animals must comply with IACUC and veterinary care rules in a properly regulated research program.

The best care of animals will impact everyone's research and keep the quality of the data at the highest level.

SOME BACKGROUND PRINCIPLES

From a larger point of view, opinions on the ethical use of animals in research vary. Many animal rights activists feel that sentient animals are equal to people and should not be used in experiments in any way. But, this view is not held by many others, including those who use animals in research. After much thought and discussion by the research community and responsible non-researchers, the use of animals in research is considered to be justified if there is benefit to humans and sometimes to the animal species used, and if pain and discomfort in the animals are minimized. Every investigator that uses animals should be able to discuss the merits of their research according to these justifications.

One guideline that is often stressed is "Replacement, Reduction and Refinement." (7) Replacement means to replace animals with another model such as cell cultures or with other animals that are lower on the phylogenetic scale. Reduction means to reduce the number of animals that are used, perhaps by reducing the number of animals that are lost, or by a better experimental design. Unnecessary experiments or the excessive use of animals cannot be condoned. Refinement means to minimize pain and distress to the animals. All researchers should be able to justify their research with animals by addressing each of these three principles.

All experiments should be fundamentally sound. Carrying our inadequately designed experiments is in itself an ethical problem.

Additional guidelines are as follows (from ref 4, p 60), and every researcher should be able to address each of these in the context of their research.

1. "Follow the rules and regulations for the transportation, care and use of animals;
2. design and perform research with consideration of relevance to human or animal health, the advancement of knowledge, or the good of society;
3. use appropriate species and the minimum number of animals to obtain valid results, and consider non-animal models;
4. avoid or minimize pain, discomfort, and distress when consistent with sound scientific practices;
5. use appropriate sedation, analgesia, or anesthesia;
6. painlessly kill animals that will suffer severe or chronic pain or distress that cannot be relieved;
7. feed and house animals appropriately and provide veterinary care as indicated;
8. assure that everyone who is responsible for the care and treatment of animals during the research is appropriately qualified and trained, and

9. defer any considered exemptions to these principles to the appropriate IACUC."

IACUC APPLICATIONS

Many Institutions have published guidelines for the care and use of animals and the submission of IACUC applications. For example, Emory University has such information online (8). It is very detailed and helpful.

While the format of IACUC applications may vary among institutions, certain key topics are common. An important part of an IACUC application is the lay language summary which describes, in easily understood language, the reasons and justifications for the project, and explains how animals will be protected from unnecessary pain or distress. The descriptions must be given in detail and readily understood by intelligent nonscientists. It should also address the three R's described above and the various other guidelines that apply.

An important part of the application has to do with the pain and discomfort experienced by the animals and how researchers plan to minimize it. They will have to clearly describe the level of pain and distress in the animals which goes from no pain to significant unrelieved pain and distress. Researchers are expected to describe in detail how they will monitor the experiments and how they will manage the particular level of pain and distress that occurs. The greater the level of pain and distress, the greater scrutiny the procedure will receive. It is possible to get approval for experiments where pain is not treated, but the justification must be compelling. Also, the conditions and methods under which animals will be euthanized must be given; this is also required in NIH grant applications.

The application will also ask for a description and record of the training of everyone who comes in contact with the animals, as well as their knowledge about managing pain, distress and euthanasia. The veterinarians and the IACUC can help researchers get proper training.

Because of the great variations in research projects and techniques, the contents of IACUC applications also vary greatly, and each animal user may need to consult their local veterinarians and Committees for guidance. Because of this diversity, the job of the IACUC is often challenging, and their needs must be respected to maintain proper animal care at every institution. While the guidelines are stringent, ground breaking research is done across this country with excellent cooperation between researchers, veterinarians and IACUCs.

VIOLATIONS

Violations of guidelines or of existing, approved protocols can be reported by anyone concerned about the welfare of the animals. This includes the laboratory staff or individuals involved in animal care. The report can be anonymous and made to a hotline. The only experiments allowed are those clearly described in an IACUC approved protocol. If research goes in unexpected directions, then an amendment to the existing protocol is required.

If complaints are made, the IACUC will investigate them and decide if they have merit. If they have merit, then some action must be taken to correct the problem. The actions can simply be training to address the issue, or, in the extreme, the action could result in temporary or possibly permanent closure of access to animals. Keys to avoiding such problems are training and communication with veterinarians and IACUC's representatives. All violations and how they are remedied are reported to OLAW and AALAC.

CASES

Case 1. Experiments without treating pain.

Your research team, using rodents, has been developing new procedures for organ transplants which involve improved surgical tools and new reagents that control bleeding and promote healing. If approved by the FDA, this is likely to majorly improve transplant technology at least for some organs. Your team is now ready to assemble all of the steps into a final test of the overall procedure in non-human primates. Unfortunately, some recent literature reports strongly suggest that opiate analgesics will interfere with the new, tested reagents in monkeys. Accordingly, to fully understand the conditions and limits of the technique, some animals must be studied without the opiate treatments for pain following surgery.

Is it possible to use animals under these conditions where the reduction of pain and suffering with opiate analgesics must be kept to a minimum or is absent?

What would you include and emphasize in the IACUC application to justify this research?

Case 2. Old reliable vs a new approach.

You routinely use an animal model of disease that requires surgery where there is some mortality of the animals. This approach is widely used and accepted as an excellent model. Recently, another laboratory has published a new procedure with reduced mortality that does not utilize surgery but rather a drug treatment paradigm that, at least according to their data, looks very good. The supervising veterinarian at your institution finds out about the new procedure and approaches your team. The veterinarian says that you need to use the new procedure because there is less pain and discomfort, and more animals survive. Your team vigorously objects and says that they don't want to risk problems and faulty data by suddenly switching to the new procedure. Also, it is not yet repeated nor widely accepted. But the vet says that your research will be shut down if you don't change. The vet cites the three R's principle as well as other guidelines.

Can your experiments involving this procedure be shut down legally?

Discuss the basis for the veterinarian's arguments.

How can you react to the situation? Is there something you can propose?

Case 3. Adherence to the approved protocol.

You have a fully approved IACUC protocol and the experiments are progressing. Some new data from your experiments as well as some from another lab suggests that a new type of drug be given to the animals with benefit. Because the new drug is similar to the old one, and because it is expected to have the same beneficial effects with fewer bad side effects, you begin to use the new drug immediately. Later, you submit a modification of the protocol to the IACUC for approval of using the new drug because you know you have to do this. What will the IACUC say about these decisions?

If you were going to purchase just one reference book about animal experimentation, what would it be?

Case 4. Be able to justify your animal work with everyone.

Your colleague has been asked to give a press conference on the latest discoveries of your team. The discoveries involve the use of rodents in the experiments. After a trial run of the conference in front of the entire research team, you did not think that the presenter had any idea why rodents were needed and how that should be justified. But everyone wants the press conference to go well.

What guidelines would you suggest that the presenter study and practice presenting? How many different ways can the use of animals be justified?

SUMMARY

The public derives great benefit from the use of animals in research, and the research can only be done if there is such benefit. Also, pain and discomfort of the animals must be minimized. As several researchers have said, the use of animals is a privilege and not a right, and the use must be according to guidelines. The ethics and guidelines of the use of animals has been studied and discussed by many for many years. Accordingly, there are a number of laws, groups and institutions and their relevant publications that regulate and guide the use of animals. In any local setting, the researcher must work with an attending veterinarian and an IACUC committee; these will help researchers become informed and meet guidelines. Researchers must also be trained and their laboratory work is likely to be inspected. As noted above, proper caring for animals not only keeps everyone in accord with humane and ethical principles, but also the best care of animals will impact everyone's research and keep the quality of their data at the highest level.

REFERENCES

1. Guide for the use and Care of Laboratory Animals (8th edition). The National Research Council of the National Academes. The National Academic Press. Wash DC. 2011.
2. Guidelines and policies for the conduct of research in the Intramural Program at NIH. NIH Office of the Director. Fifth edition, May 2016 p. 25.
3. http://grants.nih.gov/grants/olaw/tutorial/terms.htm, accessed on October 5, 2016.
4. Steneck NH. Introduction to the Responsible Conduct of Research. Revised Edition. Office of Research Integrity, 2007.
5. Guide book for IACUCs at https://grants.nih.gov/grants/olaw/GuideBook.pdf, accessed on October 6, 2016.
6. http://grants.nih.gov/grants/olaw/tutorial/terms.htm, accessed on October 17, 2016.
7. Russell WMS and Burch RL. The principles of humane animal experimental technique. London. Methuen Press. 1959.
8. http://www.iacuc.emory.edu/policies/ accessed on November 4, 2016.

4

Human Experimentation

The ability to treat human maladies and improve our quality of life depends to a large extent on human experimentation and testing. But, because of possible abuses, it has become regulated. The ethical requirements for testing in humans are designed to offer maximal protection to the human subjects and to ensure that the risks of experimentation are offset by potential gains. In that regard, the design of the experiments must be such that valid conclusions can be drawn from the results and that the study is feasible.

This chapter is a brief overview on how to approach human experimentation. An important consideration is that experimentation with humans is a serious undertaking that requires training and mentoring. Training is a key not only for avoiding problems but also for elevating everyone's research to a high level of quality.

BACKGROUND OF THE PROTECTION OF HUMAN SUBJECTS

This history is a record of some serious breaches of human rights, and the steps taken to prevent them. Briefly, the chronology of some major policies is as follows. While research on human subjects began before World War II, the codification of rules for studying human subjects occurred mainly after World War II with the notable Nuremberg Code in 1947 (1, 2). The Declaration of Helsinki (1964, with the most recent revision in 2013) also provides ethical guidelines for researchers (3). The PHS act of 1985 (4) requires that experimentation with human subjects be approved by a properly trained committee. Also, the "common rule" adopted in 1991 set the rules to be commonly followed when using humans in experiments. The common rule addresses protections of vulnerable populations such as pregnant women, their fetuses, and newborns. It also addresses the problems in using prisoners as subjects and the protections for children in research.

The important Belmont report of 1979 (5) also sets forth key principles: respect for humans, beneficence, and justice. The principle of respect implies that humans are free and autonomous and that those who have diminished autonomy must be protected. Their free choice must be respected. Beneficence is an obligation to do no harm or minimize harm, and to produce maximal benefits for human subjects. Justice refers to the idea that the burdens of research as well as the benefits are to be distributed fairly to all, and again that vulnerable populations are protected. For example, a selection process that recruited subjects from a narrow income range or age range, for sake of convenience and without specific justification, would be biased and unjust.

Institutions where NIH sponsored human research is carried out must have a Federalwide Assurance (FWA) document. The FWA commits the institution to taking responsibility for the research. Individuals carrying out nonexempt (i.e., requires IRB approval) work on human subjects with NIH funding can only carry out the work at Institutions with an FWA, unless special arrangements are made. Institutions must also establish an Institutional Review Board (IRB) that examines and must approve experimental protocols.

There are yet additional policies, offices and regulations that pertain to research on human subjects. The development of these has been a long, thoughtful, and continuing process where the protection of the variety of human subjects has been addressed. More detail on this development and history can be obtained from supplementary readings and from various websites including the NIH websites (6). A most important requirement for researchers of human subjects is to get training in the topic.

Members of research teams must be prepared to explain to the average person, as best they can, how their research benefits society, and also how the various regulations are complied with.

REQUIREMENT FOR TRAINING

As noted above, studies cannot get approved without proper training of the investigators. NIH policy (7) clearly mandates that.

> "Policy: Beginning on October 1, 2000, the NIH will require education on the protection of human research participants for all investigators submitting NIH applications for grants or proposals for contracts or receiving new or non-competing awards for research involving human subjects." (7)

Research institutions offer training, and various funding sources can make suggestions or may have their own requirements for training. Emory University's requirements, for example, are stated on the University's website (8).

> "...Prior to submitting research protocols for review and approval by the Emory University IRB, all Key Research Personnel listed on an Emory IRB submission, regardless of their position, must complete the web-based Collaborative

IRB Training Initiative (CITI) Program in the Protection of Human Subjects in Research available at http://www.citiprogram.org/. Clinical Investigators will also need to complete the training required by the Emory Office of Clinical Research."

SCOPE OF HUMAN EXPERIMENTATION

Human experimentation is any investigation or test involving living humans with the goal of obtaining data or private information. Formal definitions of "research" and "human subjects" have been given (9). While medical investigations involving new drugs or procedures are obviously human experimentation, behavioral studies such as those in psychiatry, psychology, and the social sciences are as well. The latter might involve tests of various kinds including questionnaires, performance tasks, or physiological measurements under various conditions.

Some studies are considered to have a very low level of risk. One is an experiment carried out in an educational setting that involves normal educational procedures. Another is using samples, observations or tests where subjects cannot be identified. Another is studying existing data or de-identified tissue samples available to the public. As noted above, IRB approval is needed for any kind of human experimentation. Even the kinds of studies that are "exempt," (do not require more extensive official IRB review) require that an IRB declare them to be exempt.

THE IRB AND ITS REVIEWS

The Institutional Review Board, or IRB, confidentially reviews protocols for research studies of human subjects. The membership of the IRB is described in the Common Rule. It must have at least five members that include a scientist, a non-scientist, and a person without connections to the institutions or their families. Training is critical for IRB members, and COIs must be declared.

PIs can interact with the IRB staff or members of the committee when preparing their protocol. PIs may also be asked to attend IRB meetings to clarify some issues.

Every protocol received by the IRB first gets an administrative review by the IRB staff. Some studies may not need a further scientific IRB review. Is it research? Does it involve living humans? If the study is not research and doesn't involve humans, an IRB review may not be needed. If the researcher is not sure, and it seems best to be cautious, the IRB can make this determination.

The administrative review by IRB staff and committee members determines if a project is 1) exempt from additional scientific review, 2) expedited non-exempt, or 3) needing full IRB review. This determination is based on detailed criteria and must be made by the IRB (10 - 12).

Exempt research can be, for example, educational tests, observations, interviews, or surveys. But keep in mind that the IRB must make the declaration if a project is exempt from further review.

Expedited non-exempt reviews require that there is no more than a minimal risk to the subjects (same as the risks in daily living) and that there is minimal risk if the subjects are identified, and finally that the research is not classified. The Federal Register lists categories of procedures that allow for an expedited review, and there may be additional criteria for getting a non-exempt expedited review.

If a review by the full IRB is needed, the protocol will be assigned to members, and a thorough review based on all of the criteria will be carried out.

The IRB can ask for a revision of the protocol before it can be accepted or considered further. There may be an annual re-review in some cases as well. Most protocols require some revision.

Conflicts of interest (COIs), which are described in other parts of this guide, should be disclosed to the IRBs as well as to the institution; these bear on scientific rigor, possible bias, and possibly on risks to subjects as well.

THE PROTOCOL

The protocol submitted to the IRB must contain several elements, and the committee will examine these carefully.

1. A PI must be identified who will take responsibility for the project and for implementing the protocol and applying all regulations (Before submitting the protocol to the IRB, the PI may be required to submit it elsewhere as well. For example, the department chairperson may wish to review it).
2. The investigators must be adequately trained for the project.
3. What exactly is being tested and what is the hypothesis? These must be clear and the proposal should convince the committee that new, worthwhile knowledge will be gained. The premise of the work (Chapter 2) should be stated. The standard of care and the exceptions from a standard of care should be described if applicable.
4. The project must be scientifically sound. It must be designed rigorously enough to provide sound conclusions. The number of subjects must be justified and a description of the power analysis and statistics should be described, and the experimental design should be clear and understandable by the full IRB. Carrying out inadequate studies is in itself an ethical problem.
5. What is the study population? The population must be compatible with the goals of the project, and it should

be as diverse as possible so that the results are applicable to as many as possible. The principle of justice must be upheld (see the Belmont report) so that procedures are equitable and fair. Are there vulnerable populations such as pregnant women, children, prisoners, mentally ill etc.., and are they adequately protected? Recruitment procedures should be non-coercive and the populations studied should not be feel coerced in any way (also see below).
6. The risks must be fully described. Are there procedures that produce physiologic or emotional risks? Will there be stress from the procedures? What are the frequency and magnitude of the risks?
7. Is there is a risk that privacy or confidentiality will be breached? How serious is this risk and what might be the consequences?
8. The informed consent document will be reviewed carefully. Not only the document itself is important, but how informed consent will be obtained is important. Are the informed consent procedures adequate? Every member of the research team is considered responsible for obtaining proper informed consent from the subjects.

Informed consent must be freely given and based on full disclosure of the facts about the experiment. Is there any type of subtle coercion such as paying disadvantaged subjects a lot of money? Does the protocol involve deception of the subjects and how is that dealt with?

Subjects must be told that it is an experiment and what is expected to be gained (the benefits) from it.

The probability of possible risks and discomforts must be explained.

There should be an explanation of how confidentiality will be maintained, and how injuries related to the experiment will be handled. With regard to the latter, a contact number or process should be given so that help can be obtained as needed. If relevant, possible alternative treatments must be explained.

Depending on the risk level, the subjects may be expected to sign a document that is kept on file.

There may be additional issues that need to be addressed in the consent document depending on the details of the study. For example, if the study involves children who are not of the age where they can give legal consent, they still must give assent to their involvement.

Exculpatory language is not allowed. This occurs, for example, when a subject waives his/her right to remuneration from the medical center if they are injured during the experiment. The subject cannot agree to no payment for injuries occurring during the study.

The subjects must be informed of their right to withdraw from the study without penalty.

Note that the informed consent document should clearly describe the study. To get an overview of the project, an IRB reviewer could read the consent form and the lay language summary first. These should be prepared with that in mind.

9. In projects where unexpected risks might develop, there must be a data and safety monitoring plan. The protocol must explain who will do it and how it will be done. How

will unexpected risks be managed? Should the subjects be re-consented and the consent document be revised? Everything to ensure safety must be done.

Note that someone trained in medical ethics, a bioethicist or a research ethicist or a clinical ethicist, is an invaluable consultant in writing protocols and carrying out human experimentation. Including such individuals in research teams is highly desirable.

CONTINUING RESPONSIBILITIES

Once the protocol has been approved, the PI and the team must adhere to the protocol and the rules for human experimentation. Any changes in the protocol needed for any reason must be approved before they can be implemented. For example, only subjects described in the protocol may be recruited. Unexpected risks to the subjects must be thoroughly considered and reported before going forward.

If the study is not carried out as described in the approved protocol, or if there is unanticipated danger to human subjects, the study may be suspended or terminated depending on the issues. This may result in a revision of procedures, notification of previous subjects, or yet other actions.

CLINICALTRIALS.GOV

ClinicalTrials.gov is an online registry and database that contains information about medical studies and their human volunteers (13). It is federally supported. Registration at the site and adding various reports are the responsibility of the PIs on the studies. While this site is very useful in keeping abreast of human experimentation in various fields, a listing on it does not reflect endorsement or approval by the NIH.

Regulations regarding the site have recently been expanded in a document referred to as the "final rule," which was put in place in 2017. It indicates which trials should be registered on this site and how results should be entered and updated. Clinical researchers must be familiar and comply with this document (14 - 16) to be in full compliance with regulations.

CASES

Case 1. Consequences of limiting your responsibility and focus.

Suppose you are a microbiologist, and you discover a protein that is capable of slowing cell growth in vitro in such a way that it could be a new treatment for some cancers. Your extensive data are convincing enough that you acquire a patent and interested parties are planning a human trial. As the protocol is being written, you neglect opportunities to learn about informed consent and you claim that that is not your responsibility because you are not a clinical researcher. You are on the protocol to deal with issues of the protein under study, and not the human side of the work.

Can there be a consequence of your refusal to be involved in the other aspects of the protocol?

Case 2. A new protocol: When do I need help and where do I find it?

You are a young and well trained investigator whose research is going well. Soon after your last interesting publication, you are writing a new research protocol to extend your work. But this extension takes you into some new and unfamiliar territory. Thus, you aren't sure how to write and carry out certain aspects of your new study. What are your options to get the new material written in a rigorous, responsible, and appropriate way?

Case 3. An unexpected threat to subjects.

You are an investigator on an approved protocol, and while you are watching just the second subject in the study, you notice a cardiovascular abnormality occurring right after administration of the test substance. You realize that the abnormality could be a serious problem and you become alarmed. You stop the test, and ask an attending nurse and physician to watch the subject for possible cardiovascular problems. You go back to the files and examine the previous research results in animals and humans, but the abnormality was never reported. You aren't sure what to do or think. Could this problem occur in only a subgroup of human subjects? Perhaps it is not even related to the test substance at all and is only a random occurrence. If the study cannot proceed, you are very concerned that an important treatment will be undiscovered and lost.

What must you do? To whom do you go for advice?

Case 4. Justifying the use of human subjects instead of animals.

You submit a proposal to the IRB for an experiment using living human subjects. But the IRB communicates that it is not convinced that the experiment shouldn't be done in animals.

In general, how do you justify using human subjects instead of animals.

Case 5. An ethical dilemma.

You are preparing a protocol for a study involving human subjects. Unfortunately, all of the tests that would be useful for testing your hypothesis have significant side effects and risks, and you aren't sure how to proceed. Your fellow researchers are also unsure. You feel that you have an ethical dilemma.

How do you handle such ethical dilemmas? Whom do you ask for help?

SUMMARY

The ability to treat human maladies and improve our quality of life often depends on human experimentation and testing. Such experimentation is a serious undertaking that will require training, mentoring, and compliance with a number of rules. Training is a key not only for avoiding problems but also for elevating research to a high level of quality. The written research protocol must clearly adhere to established rules and ethical and scientific standards, and be approved by a qualified IRB. Consultations with medical ethicists are valuable. Risks to humans must be minimized, and informed consent must be obtained and documented. Approved protocols must be strictly adhered to and any changes will require IRB approval. Violations of approved protocols (noncompliance), or the occurrence of unanticipated risks to subjects are unacceptable and must be corrected and reported. Clinicaltrials.gov is an important website where certain trials must be registered and updated.

REFERENCES

1. https://history.nih.gov/research/downloads/nuremberg.pdf, accessed on January 27, 2017.
2. http://www.cirp.org/library/ethics/nuremberg/, accessed on January 27, 2017.
3. http://www.who.int/bulletin/archives/79(4)373.pdf, accessed on March 27, 2017.
4. https://grants.nih.gov/grants/olaw/references/phspolicylabanimals.pdf, accessed on January 27, 2017.
5. https://www.hhs.gov/ohrp/regulations-and-policy/belmont-report/, accessed on January 27 2017.
6. https://humansubjects.nih.gov/ethical-guidelines-regulations, accessed on January 12, 2016.
7. https://grants.nih.gov/grants/guide/notice-files/NOT-OD-00-039.html, accessed on January 13, 2017.
8. http://www.irb.emory.edu/training/, accessed on January 13, 2016.
9. https://humansubjects.nih.gov/glossary, accessed on August 15, 2017.
10. http://www.research.uci.edu/compliance/human-research-protections/researchers/levels-of-review.html, accessed on March 1, 2017.
11. https://www.hhs.gov/ohrp/regulations-and-policy/guidance/categories-of-research-expedited-review-procedure-1998/, accessed on March 1, 2017.
12. https://research.sfsu.edu/protocol/review_process/review_categories, accessed on March 1, 2017.
13. https://clinicaltrials.gov/ct2/about-site/background, accessed on August 15, 2017.

14. https://www.federalregister.gov/documents/2016/09/21/2016-22129/clinical-trials-registration-and-results-information-submission, accessed on November 5, 2017.
15. https://prsinfo.clinicaltrials.gov/, accessed on November 5, 2017.
16. Deborah A. Zarin, M.D., Tony Tse, Ph.D., Rebecca J. Williams, Pharm.D., M.P.H., and Sarah Carr, B.A. Trial Reporting in ClinicalTrials.gov - The Final Rule. N Engl J Med 2016; 375:1998-2004November 17, 2016DOI: 10.1056/NEJMsr1611785.

5

Data Acquisition, Management, Sharing and Ownership

Data is the essential product of the research scientist. Everything in actual research is built on data. Therefore, how to manage data is of the utmost importance. Having discussions with mentors, collaborators and colleagues will be helpful because they have useful knowledge and experience. Funding agencies and institutions also have polices on data (1-5) and these should be examined by all. Aside from specific requirements, practices for managing data may vary, but the more extensive recommendations in this chapter are likely to prevent various problems for the investigators.

THE NOTEBOOK

Just about everyone that reaches graduate school has been instructed about the use of notebooks. Notebooks are the depository of everything about a researcher's experiments and their data. They are vitally important, can be legal documents, and therefore several rules and procedures are required for them. They may help decide about giving fair credit, document support for patents, or may help someone to repeat the experiments in future years (6). Keeping good records on data may require patience and care.

First, everyone must be familiar with the requirements and practices of their laboratory, department and institution (1-6) that may impact on managing data. But there are many general and widely accepted practices and rules.

1. The physical notebook is traditionally considered to be a hard bound and page numbered book. But considerations need to be made for supplemental storage of data and items than cannot be incorporated into that format. For example, western blots, scans, images, lab test results, large computer analyses and the like also need to be kept. Thus, the notebook must be the guide to where and how additional data are stored. For similar reasons, some feel that binders are more flexible and useful as notebooks because of the ease of including various data types. There are also electronic notebooks available as well. Thus, the word "notebook" can imply many formats of record keeping, although traditionally it is a hardbound and page numbered book. Specific labs may have specific requirements.
2. Each lab should have a manual of standard procedures that focuses on its specific interests and protocols, and references to those can be included in the notebooks. This is convenient and can save on entries, and such manuals should be kept with the notebooks.
3. Organize notebooks conveniently. It may be convenient to use a different notebook for each different project or publication. A detailed table of contents is essential for navigating the pages and finding topics easily. The notebook is the complete guide to a researcher's experimentation.
4. Notes should be made in ink, and changes such as crossing out or adding information must be initialed and dated

and the reason for the change should be given. In fact, everything must be dated and the author clearly identified.
5. Notebooks never leave the lab, but copies can if the supervisor agrees. If someone wants to work at home, only copies may be taken. Note that in clinical studies, the inclusion of patient identifiers in notebooks create special problems; be sure that copying such data is allowed and be sure everyone's responsibilities for the copies is understood.
6. Notebooks are so important that many feel it is important to have another (mentor/adviser) examine, sign and date the books regularly. In this spirit, copies of the books should be made at regular intervals and the copies stored safely elsewhere. Keeping data records in more than one format (hard copy notebook and electronic) and place is a sound practice. If patent work is being done, then cosigning, dating and multiple storage sites are essential. Note that many institutions have explicit guidelines for data management (4, 5).
7. How long should notebooks and data be kept? Answers range from 3 years to forever, and the correct answer for the researcher will likely depend on the funding source and the policies of the institution! The most frequent answer may be about 7 years after the end of the project, and this is an NIH recommendation (1). If the data support patents, then many years of storage are required, and exact requirements can be gotten from the patent officer. It is prudent to inquire about storage time and practices, and to plan on keeping data for many years.
8. A different kind of notebook, which is not really a research notebook, is the personal scientific diary. These personal

diaries can include new ideas, discussion summaries, things to look up and so forth. While the diary is not a laboratory notebook, it can be a valuable personal companion.

WHAT GOES INTO THE NOTEBOOK?

Essentially everything about the experiments should be included with dated entries. Simply put, the notebook should tell the reader what, why, by whom, when, how, what it means, and so what.

1. A title or brief description should begin each new experiment. It should be entered in the table of contents.
2. A clear description of the hypothesis, its premise and rationale for each experiment should be given. Also, this would include references to key publications that impact the hypothesis. It could include a justification for the techniques that are used, the number of animals, the statistical analysis, and so forth. The scientific premise (Chapter 2) underlying the work is important.

    Note that the language of the notebook is in the language of the host country. If a foreigner is working in the USA, for example, then it may require extra effort to learn the language and write English well. It must be legible and understandable.
3. A description of the procedures and protocols is needed. Note the chapter on rigor and transparency. Any colleague should be able to repeat the work from the notebook.
4. The raw data or a description of its whereabouts and details on how to access it should be included.
5. The analysis of the data should be clear. Print outs of statistical analyses should be kept.

6. The conclusions of the analysis and its importance are needed. If the experiment is considered a failure, state why. Are new directions needed? If successful, then state why and the implications.
7. Everything else about the experiment should be included. These might be someone's suggestions, IACUC approval numbers, controlled substances records, etc., or their locations.
8. As noted above, if changes are made or if items are deleted or added, then the changes must be explained, initialed and dated.
9. When in doubt, or if someone has a question about notebooks and data, advice can be gotten from more experienced investigators, or official guidelines (1-6).

WHO OWNS THE DATA AND WHO IS RESPONSIBLE FOR IT?

The experimenter should be able to answer this key question before he/she begins so that there are no misunderstandings about how the data can be handled and used. The questions are: Who owns the data? What are the requirements for its handling? What rights do I have, and can I publish it?

Regarding ownership, it may depend on where the funds come from. In the US, government agencies that provide grants, for example, assign ownership of data to the research institution that handles the funds and employs the investigators. If the funds come from industry, the company may seek to retain ownership and control of the data, but the researcher's institution may have a policy that affects that. Other organizations may do it differently and assign ownership in different ways. With government funding, there is also a distinction between grants and contracts. Grants allow the PI to use the funds and the University owns the data, whereas with contracts there is usually a specific deliverable and usually the government owns the product. Where are you in this scheme?

With government grant funds, the data are not owned by the investigator as noted above. Funding for research is usually given to the institution and not the PI. Hence, the University owns the data. The PI, who has a very vested interest in the data, is the custodian of the data and may use it in the most effective way. The PI is also the custodian in the sense that he/she is responsible for its storage and access. Always act as though the data are owned by someone else. Some institutions require notification and approval if data are going to be destroyed.

If a researcher leaves his/her institution, they cannot automatically take the data with them. The current institution may have obligations regarding the data. But copies of the data can usually be taken with permission. It would be best to discuss this with the appropriate officials at the present institution and also with those at the new destination.

THE BAYH-DOLE ACT FOR PATENT ISSUES

The Bayh-Dole act is legislation that addresses the use of data for patents that derive from federally funded research (7, 8). It allows inventions to be held by small businesses, universities, or other non-profit institutions. The PI can be assigned an inventor role and may accept some profits from such inventions; the institution usually has a policy about how much benefit investigators may receive. If the work seems patentable with a good possibility of income, then this should be discussed with the patent officer at the institution. The manner in which data are acquired, processed, and stored may be impacted by this.

CONFIDENTIALITY

Before one starts a project, they need to know how much confidentiality is expected. The degree of confidentiality depends on the sources of funding, the nature of the funding (grants or contracts?), the policies of the lab, supervisor and institution, and the reliability of the data at some point in time. No one would tell about the findings unless they were solid and repeatable. Many feel that findings are confidential until fully approved by peer review. In this case, nothing is discussed with others until the paper is formally accepted for publication. A casual mention of someone's unpublished work to others could result in confusion about priority.

Confidentiality must be balanced by the need to get advice from others, such as the useful practice of discussing data at group meetings. This works both ways. Expect confidentiality and give confidentiality to others.

SHARING DATA

There is a general expectation that raw data and related material should be shared at some point. This promotes transparency and efficiency in research and justifies the use of public funds. Obviously everyone shouldn't repeat and reproduce the same data. Only some groups might have the ability (equipment, reagents) to produce certain kinds of data. So sharing data makes sense. But certain circumstances may impact this view. Certainly, only solid, repeatable data should be shared whereas preliminary findings that require repetition and additional processing probably shouldn't. If the data impacts on a current public health problem, then the data become more important and should be shared when the PIs are reasonably certain of the findings. Data from NIH supported clinical trials are reported on clinicaltrials.gov. Normally, the publication of a paper by researchers is considered adequate sharing, but it is possible that the availability of raw data prior to processing may be of interest to others. Many journals require sharing of raw data and other information about the experiments after publication. There is also the Freedom of Information Act (FOIA) that requires sharing of information related to federally funded research (9, 10). Thus, PIs must be prepared to share data and related information. Also, because they may not own the data, approval may be needed for any sharing. Records of this process should be kept.

BIG DATA

Big Data refers to data sets that are extremely large and/or complex. The challenge facing experimenters is how to collect and store it, and how to effectively analyze it. There are other issues such as how to share it.

While the history of this problem goes back many decades, it has been more recently brought into the life sciences because certain techniques and approaches produce massive data banks. Genome sequencing can produce gigabytes of data, as can brain imaging experiments. These data banks obviously cannot be entered into notebooks, but the notebooks still need to be transparent and complete with information on how to access all stored data. If someone finds themselves with "Big Data" problems, then they need to consult an expert in their field as the approach may depend on the type of experiment and discipline.

CASES

Case 1. Can I destroy older data?

Dr. Black, a colleague of retired Dr. Gray, finds the records and stored files of Dr. Gray in a university storage area. The files contain old grant applications, published papers, correspondence, and collections of original data. Dr. Black, without thinking very much, decides to put the files of the retired Dr. Grey in the trash to make room for other things. It doesn't seem like the files will be needed.

This simple scenario brings up several key questions. Who owns the data in the files? Who is responsible for its storage and access? Is permission needed to examine the files of a colleague, particularly if there is unpublished data? Under what conditions can the data be discarded and put in the trash? What do Dr. Black and other investigators need to know? Supposing Dr. Gray doesn't care about the old files? Does that change anything?

Case 2. Painful consequences of careless record keeping.
Mary, a new student working for Dr. Paul, has recently started doing experiments and using her new lab notebook. She is inexperienced, so her notebook isn't as complete as it should be; for example, her experimental descriptions are a little too short. Much of her data are large spreadsheets which she stores on her computer. As the project nears completion, and as she gets ready to produce a manuscript, she finds she is hopelessly disorganized. She goes to Dr Paul, and she is totally contrite. But, she can't find key data partly because the experiments were not described well and the notebook is disorganized and incomplete. Some of the files on the computer cannot be found presumably because they were not titled and stored in an organized way.

What can be done? Who is at fault – Dr Paul or Mary? How should Mary's behavior and notebook practices change? Should the notebook contain information on the location of stored files in computers?

Case 3. Handling widely distributed data.
In a multidisciplinary or highly collaborative project, data can be scattered over many labs, notebooks, computers, etc. Who should be in charge of getting copies, noting where the data reside, and how they are accessed? How should it be organized? How often?

Case 4. A template for describing experiments.
You are a new dissertation student entering the lab of a new faculty member. You are excited because the new faculty person brings many new techniques and you are excited to be learning them. As you begin experimenting, you see that different lab members are using slightly different formats for recording their experiments in notebooks. While there is nothing wrong with this, you notice that details are sometimes left out; these include dates, or sources of reagents, or some procedures. In previous training, you saw that a uniform template for describing experiments was helpful and reduced these errors. Accordingly, you ask your advisor if you could prepare such a template for this lab. He enthusiastically says yes.

How would you prepare such a template for lab notebooks? What would it contain?

SUMMARY

Data is the basic product of the research scientist. It is the foundation for new knowledge and translational advances. Accordingly, there are good practices, data management plans, and regulations regarding the collection, storage, processing, sharing and ownership of data. The notebook and its extensions (data storage and processing platforms) are the basic places where data are kept as part of a detailed description of the project. Notebooks are very special in that they do not leave the laboratory, may be signed and dated, and multiple copies are stored in multiple places. Depending on how the data are used, they must be kept for many years. Data ownership may depend on the source of the funds, but federal funds are given to the Institution and the Institution owns that data. The investigator is the manager of the data and responsible for how and where it is stored. Depending on the situation, data can be kept confidential, and there are some expectations that data will be ultimately shared as well. Every scientist is expected to fully understand what is required of him or her with regard to data.

REFERENCES

1. Guidelines for Scientific Record keeping in the Intramural Research Program at the NIH. https://oir.nih.gov/sites/default/files/uploads/sourcebook/documents/ethical_conduct/guidelines-scientific_recordkeeping.pdf, accessed on September 14, 2016.
2. Guidelines and Policies for the Conduct of Research Program at the NIH. https://oir.nih.gov/sites/default/files/uploads/sourcebook/documents/ethical_conduct/guidelines-conduct_research.pdf, accessed on September 14, 2016. Section on Data management and archiving.
3. Bird SJ, Giffels J, Golembewski E, and Vollmer SH (special volume editors). Responsible Data Management. A special issue of Science and Engineering Ethics, Vol 16 (4), 2010.
4. Polices on data management at Emory University: http://web.library.emory.edu/research-learning/scholcomm-datamgmt/research-data-management.html, accessed on November 3, 2016.
5. http://guides.main.library.emory.edu/datamgmt, accessed on November 3, 2016.
6. https://www.training.nih.gov/assets/Lab_Notebook_508_(new).pdf, accessed on November 3, 2016.
7. https://grants.nih.gov/grants/bayh-dole.htm, accessed on November 3, 2016.
8. https://grants.nih.gov/policy/intell-property.htm, accessed on November 3, 2016.
9. https://www.foia.gov/, accessed on November 3, 2106.

10. https://www.nih.gov/institutes-nih/nih-office-director/office-communications-public-liaison/freedom-information-act-office, accessed on November 3, 2016.
11. www.onlineethics.org, accessed on September 15, 2016.

6

Authorship

Publications are units of work, productivity, and progress in the research world. They are the proof of a researcher's efforts and abilities. Also, it is of no minor importance that everyone's personal authorship record is critical for advancement. Research careers are usually built on authoring peer reviewed publications. Due to the importance of authorship and also because of the occurrence of multiple authors, conflicts and disputes can sometimes occur. Accordingly, there are ethical guidelines for determining authorship.

WHAT CONSTITUTES AUTHORSHIP?

The International Committee of Medical Journal Editors' (icmje.org) offers four criteria for authorship.

1. "Substantial contributions to the conception or design of the work; or the acquisition, analysis, or interpretation of data for the work; AND
2. Drafting the work or revising it critically for important intellectual content; AND
3. Final approval of the version to be published; AND

4. Agreement to be accountable for all aspects of the work in ensuring that questions related to the accuracy or integrity of any part of the work are appropriately investigated and resolved." (1)

Most journals refer to the ICMJE or similar criteria for authorship. The Journal of Drug and Alcohol Research says "…requirements for authorship (a significant intellectual contribution, a role in the writing and editing of the manuscript, and final approval of the submitted manuscript)" (2).

The Proceedings of the National Academy describes requirements for authorship. In part, it says that "**Authorship** must be limited to those who have contributed substantially to the work. The corresponding author must have obtained permission from all authors for the submission of each version of the paper and for any change in authorship… all collaborators should have in place an appropriate process for reviewing the accuracy of the reported results. Authors must indicate their specific contributions to the published work." Additional comments are given (3).

Because situations can vary, and if there is some doubt as to whether authorship is warranted or what criteria should be used, a researcher should consult with more experienced scientists in their field or with the instructions for authors for the journal under consideration, or perhaps even with their funding source. The NIH website includes information on authorship (4).

Also, authors should be able to connect the various coauthors with their specific contributions to the paper. If contributors do not meet the criteria for authorship, then they should be noted in acknowledgments and their specific contribution could be stated.

MULTIPLE AUTHORS AND POSITIONS IN THE LIST

Over the past decades, the number of authors per paper has been increasing. Different groups have different policies about authors' positions in the author list. A common arrangement is that the last author is the overall group or team leader who is often the most senior of the authors. But this is not always the case. The first author is often the leader of the specific project described in the paper. This person often does most of the work, and works closely with the supervisor or lab leader so that data acquisition, storage and analysis are carried out properly. The first author usually leads in the writing of the paper and works with the supervisor and coauthors so that the paper accurately reflects the work and findings. Usually all of the papers deriving from a trainee's project will have the trainee as first author. "Middle" authors can be arranged in the order of decreasing work or contributions to the paper.

Sometimes, two of the authors contribute equally. If allowed by the journal, that can be stated with a footnote in the paper's title and author line, and on the author's curriculum vitae. Before the paper is submitted all authors need to agree that the authorship line and their contributions are accurately stated.

CORRESPONDING AUTHOR

The corresponding author takes responsibility for submitting the paper along with required forms. He/she deals with the correspondence with the journal and other coauthors during the submission and reviewing process. Sharing all correspondence with coauthors is recommended. Communications from colleagues may also be sent to the corresponding author.

DEALING WITH DISPUTES

The authorship list and the tasks assigned to coauthors should be clearly decided in advance of starting the research, and the discussion should continue until everyone involved openly expresses agreement. If situations and work assignments change, and they often do, the project leader needs to make relevant changes in authorship and also make these changes known to the group.

MISCONDUCT

If one of the authors on a paper is charged with misconduct such as plagiarism or fabrication, then all of the authors are affected in some way. It is possible that the paper will have to be withdrawn and innocent authors may lose their work and may even be tainted by the accusation. While the senior author is often held responsible for the integrity of the papers, the other authors have a vested interest as well. In so far as misconduct can be prevented, everyone involved in the project should be aware that bad things can happen to the innocent due to the actions of a single person.

CONFLICTS OF INTEREST

The submitted and published paper should clearly state all COIs that any of the authors have. More on this is given in Chapter 9.

CASES

Case 1. Can the agreed upon order of authors change?
Edward, a new graduate student, enters the lab of Dr. Gold to do his dissertation research. Before starting his main project, Dr Gold asks Edward to finish up the project of a student who has left the lab for a fellowship opportunity. Dr Gold explains that very little

needs to be done, but if Edward will do the work, he will be the second author on the resulting paper. Edward feels that this is an excellent opportunity to learn something new and to become a co-author on a publication, so he agrees. But, as he does the needed experiments, he finds that he cannot repeat the work of the earlier student. After some investigating, it seems that one of the borrowed reagents is at fault. Further, the source of the reagent agrees that the reagent was bad but that they never realized it. Accordingly, Edwards finds that he has to repeat most of the work and a much greater effort will be needed. He does complete the work and the paper can now be written. Edward now feels that he should be first author since he did almost all of the published work.

What should Edward do and how should he proceed?

The earlier student whose work can't be repeated feels that he/she did the work, made key decisions about the direction of the work, and still should be first author. Do you agree? How should this be resolved?

Who should make the decisions?

Case 2. Who can be an author?

Thomas is a graduate student in the lab of Dr. Rome where he is doing nice work on a new protein involved in neuronal plasticity. A very skilled and experienced technician works with Thomas and purifies the protein which is a major step in the project. The ensuing draft of the publication included all of the contributors as co-authors, but the technician's name was not included as a coauthor but was in the acknowledgments.

The technician complains and says that his/her contribution was unique and important because no other collaborator had the skills to do it. Moreover, many significant findings derived from the purification work.

How do Dr Rome and the collaborators resolve this?

What decision would be made if the ICMJE criteria for authorship were used?

Case 3.

You are a younger scientist and you have been invited to contribute to a project. You are very pleased because if you do the prescribed work, the group leader says that you will be a coauthor on the resulting paper. You really want to do this and get credit for your work, so you explore what is necessary to be a coauthor on a research paper. What are the things that you are required to do to be worthy of co-authorship?

SUMMARY

Authoring papers is the standard by which researchers are judged. Because of its importance, the criteria for authorship have been extensively discussed and outlined. Senior team leaders or investigators should discuss with all collaborators, both their role in the project and their position in the list of authors, before the project begins. If changes need to be made later, it should be made known to all involved. Any possible conflicts of interest or commitment should be clearly stated in the published paper. Usually the first author or the senior leader will be the corresponding author with the journal editors. Authorship disputes need to be managed, usually by the senior project leader.

REFERENCES

1. http://www.icmje.org, accessed on Dec 8, 2016.
2. http://www.ashdin.com/journals/jdar/guidelines/, accessed on Dec 8, 2016.
3. http://www.pnas.org/site/authors/editorialpolicies.xhtml#iii, accessed on April 4, 2017
4. https://search.nih.gov/search?utf8=%E2%9C%93&affiliate=nih&query=authorship&commit=Search, accessed on Dec 8, 2016.

7

Peer Review

One of the most important components of the scientific process is peer review. Peer review is the critical examination, by a peer, of a submitted document with the goal of providing an expert opinion on its quality. If the document is a paper or a grant, the review will impact on whether or not the paper is published or whether or not the grant is funded. It is not a small thing. This chapter will focus mostly on publications and grant applications, but the concept of peer review can be extended to many other situations as well.

Considering scientific articles, editors of journals bear a significant responsibility for a sound peer review. They must find multiple, suitable peer reviewers, have explicit requirements for submitted papers, have a system in place that handles such reviews and makes judgments about the submitted work. Similar comments apply to mangers handling grant reviews

The NIH offers guidance on grant reviews (1), and various journals offer guidelines for reviews of papers as well (2-4).

PURPOSE OF PEER REVIEW

Peer review serves the community, editors and authors by assessing whether a document is worthy. In the case of a submitted paper, the reviewer will judge if the submission is new, rigorously carried out, and interesting. It is in the best interest of the authors to emphasize these elements in their writing. The reviewer, based on his/her overall impression, will submit a score or judgment as to whether or not the paper should be rejected, published, or possibly revised and then published. In the case of a grant application, the review will assist the manager in judging the quality of the proposal and how it ranks against other applications; peer review is a bedrock of grant funding and publishing.

The reviews help the authors because the reviews can guide the improvement of the document. Incorporating the reviewer's comments or suggestions into the revised documents is commonly done.

REQUIREMENTS FOR SOUND PEER REVIEW.

1. Peer reviewers must be qualified for the task. They must have both the knowledge and experience to make judgments. If they do not, then they should decline the offer to do the review. If reviewers need to include colleagues for a thorough review, then it seems best to ask the editor if that is acceptable, but before the review. More junior people, even trainees, can be qualified to review in some cases.
2. Do the potential reviewers have conflicts of interest (COIs)? COIs must be declared to editors or funding agencies, and there are many types of COIs (Chapter 9). One might occur when the writer of the grant or paper is a former student of the reviewer. Another occurs when reviewers are asked to critique the work of a faculty member at the same university. If the potential reviewer is a recent coauthor (say within 3 years), or if the topic of the paper or grant is similar to one they are writing, then they must declare that and perhaps excuse themselves. Reviewers might not be appropriate if

they object to the work on religious grounds, or if there is an ongoing competition between the reviewer and the author. A financial COI occurs when the reviewer stands to gain or lose financially depending on the outcome of the review. Reviewers must be independent and not associated with the Journal as well. Even if there is simply an appearance of a conflict, then they may wish to decline. If they have COIs but still feel that they can do a fair review, then it should be discussed with the journal editor or grant manager.
3. Reviews must be confidential. If there is a reason why that cannot be done, then it must be shared with the editors and the reviewers should disqualify themselves. The document may contain key ideas that will impact the field; these belong to the authors and they need to get the credit for those. Reviewers cannot use any information from a reviewed document until it is publicly available. Reviewers cannot share the information with anyone else without a good reason and permission of the editor.
4. The review must be completed in a reasonable time; sometimes a deadline is given. If the deadline can't be met, then the potential reviewers should decline the review. In a competitive field, timeliness is important and not meeting the deadline is unfair to the authors.
5. The review is expected to be thorough. For example, is the paper suitable for the journal? Is the premise given? Is the methodology sound and transparent? Are there problems with the conclusions? Are they supported by the data? Has the grant or paper been reviewed according to all the criteria? Etc.
6. An important consideration is the writing. Is the document understandable? If not, then it is not acceptable.

7. Do possible reviewers have a bias beyond what was discussed in the COI section above? Editors need to select reviewers to avoid biases as best they can.
8. Reviewers must address ethical issues. Are animals in the study given proper protection? Are the rules for using human subjects followed? Are COIs or lack thereof stated? Does the methodology as written make the experiments transparent and easily repeatable (chapter 2)?
9. It is most useful to write a review that is helpful and constructive for the authors. Has a possible interpretation been missed? Are the authors relying too much on certain data? Are the calculations and statistics done properly? Is the methodology transparent? Have they missed an important citation that will impact on their study? In general, it is helpful to explain the basis for any of the reviewer's criticisms.
10. In keeping with confidentiality, the manuscript should not be kept where it may be seen by others, and it should be destroyed after the review. The journal or funding agency is likely to have a policy on that.

PROBLEMS WITH PEER REVIEW

Peer review is not perfect as all experienced scientists know.

1. While there are computer programs that can identify improper manipulation of images or plagiarism, an unpleasant fact of life is that reviewers, editors and the like cannot detect all fraud or misconduct.
2. Sometimes reviewers and editors don't do their jobs well enough. Directions are not supplied or read, and the reviews are not timely or thorough.

3. It can be difficult for editors to find qualified reviewers who are willing to give up their time to do a review; reviewers are rarely paid or may be given only a token honorarium. The need for multiple reviewers can add to this difficulty.
4. Because multiple reviews are the norm, recommendations of more knowledgeable reviewers may be over ridden by a majority of other opinions from less knowledgeable reviewers.
5. Biases are difficult to control but do exist. Some studies suggest that reviewers are affected by the status of the authors or institutions that produce the papers. Biases can be generated by personal feuds, differences in professional background, gender and other factors. Biases may be especially impactful in controversial research that is new in direction, or with hypotheses that are very innovative.

There have been suggestions to improve the peer review process. One is to use a double blind system where the reviewers are blind to the author and the authors are blind to the reviewers as well. But, one could argue that knowing the authors helps in assessing relative skill and credibility. Another is the use of unblinded reviews where both the authors and reviewers are known to all. Yet another possibility is for the journal to publish all the review documents along with the final paper so that readers can see the issues raised by reviewers and the responses of the authors (5); this transparency makes everything open. There are relative advantages and disadvantages to these various ideas.

In any case, in spite of problems with confidential peer review, it is still considered the bedrock of the review process. Instead of changing the process in major ways, efforts are continually being made to improve the existing peer review process.

CASES

Case 1. Am I a suitable reviewer?
You are asked by a journal editor to review the work of a colleague in your field. What questions must you ask yourself in order to decide if you are a suitable reviewer? Supposing there are one or two items that you are unsure about, who can you ask for clarification?

Case 2. Confidential information that will impact you.
Suppose that a journal editor sends you an article to review and you see the entirety of the paper. Then you realize that this paper is significantly similar to something you are currently preparing for publication. You are now aware, given the knowledge of the article to be reviewed, that you could improve your work by making changes mentioned in the reviewed paper. You want to use the new, unpublished information to improve your work. What can you do? When can you do it?

Case 3. Feel like a useless reviewer?
You have submitted a review of a paper pointing out several key criticisms that, if dealt with, would improve the paper. As it is written, it seems to be misleading, and you point this out in your review. But when the paper appears in print, the suggestions have not been taken. If you choose to act, what could you do?

Case 4. Whom do you ask about reviewers?
You are reviewing a submitted paper, and you realize that a trainee in the lab has the experience to contribute to a part of the review. Can you simply ask the trainee or how should you proceed?

Case 5. Major impact of confidential information.
You are a successful professor and you started a small company some years ago for developing novel antidepressants. You get a paper to review and, lo and behold, it just happens to focus on some work that your company is also doing. Moreover, it shows that one of your company's main approaches won't work. If that is so, the company will have spent a great deal of money without benefit and also affected the careers of many people. While you can't be sure that the paper you are reviewing is correct, you can't be sure that it is not. Ethically, you know that you cannot reveal the contents of the paper until the paper is publicly available, which could be many months or more. But because of your worries about the company, staff and money, you feel ethically bound to do something. What can you do in this ethical dilemma, if anything?

SUMMARY
Peer review assists the scientific community, editors, authors, grants managers and employers by providing an assessment, by qualified peers, of a submitted paper or grant. Reviewers have substantial power, and in some cases may influence a field of study; substantial responsibility goes along with this power. The peer reviewer must be qualified to judge impartially and ethically, able to complete the review in a timely and thorough fashion, and be without COIs. Minor conflicts may be excused by the editor or grants manager. There are additional requirement for a fair review as well. In spite of known problems with the peer review process, it is still a cornerstone of evaluation. Well written reviews can be very helpful to the submitter, and this is important. When someone is asked to be a reviewer, they should get a thorough understanding of what is asked by the person or editor, and they must adhere to ethical requirements in the review process.

REFERENCES.

1. https://grants.nih.gov/grants/peer/becoming_peer_reviewer.htm, accessed on January 6, 2017.
2. https://www.elsevier.com/reviewers/what-is-peer-review, accessed on January 7, 2016.
3. http://www.ashdin.com/journals/jdar/reviewerguidelines/, accessed on January 7, 2017.
4. http://www.pnas.org/site/authors/reviewers.xhtml, accessed on April 6, 2017.
5. http://emboj.embopress.org/authorguide, accessed on May 15, 2017.

8

Research Misconduct

Research misconduct is proposing, carrying out or publishing research in a dishonest way. This chapter focuses on the definition of research misconduct, the procedures for investigating and identifying it, and the procedures designed to protect those who "blow the whistle" on such misconduct.

DEFINITIONS OF MISCONDUCT IN RESEARCH

After careful and lengthy deliberation, the Public Health Service (PHS) has defined research misconduct as "fabrication, falsification, or plagiarism in proposing, performing or reviewing research, or in reporting research results…"(1) It does not include honest error or differences of opinion.

"Fabrication is making up data or results and recording and reporting them.

Falsification is manipulating research materials, equipment, or processes, or changing or omitting data or results such that the research is not accurately represented in the research record.

Plagiarism is the appropriation of another person's ideas, processes, results or words without giving appropriate credit. It does not include the limited use of identical or nearly identical phrases that are not substantially misleading or of great significance, and it does not include disputes among former colleagues." (1)

IDENTIFYING AND INVESTIGATING MISCONDUCT

Institutions that receive funding from the PHS are required to have procedures in place for dealing with charges of misconduct. There should be an easily found description of how a formal investigation would be conducted, the identification of an individual who will judge the merit of the accusation, procedures towards sanctions, OR the procedures for vindication of the person accused. The findings should be reported to the Office or Research Integrity (ORI).

The requirements for a finding of misconduct have been stated (1):

" 1. There must be a significant departure from accepted practices of the relevant research community;
2. the misconduct be committed intentionally, knowingly, or recklessly; and
3. the allegation be proven by a preponderance of the evidence." Note that there is a six year statute of limitations on reporting misconduct.

If there is a finding that misconduct has occurred, then there will be sanctions. For example (1): "If an individual involved in NIH funded research is found to have committed research misconduct, the administrative actions PHS/HHS may take against them include, but are not limited to:

1. debarment from eligibility to receive Federal funds for grants and contracts,

2. prohibition from service on PHS advisory committees, peer review committees, or as consultants,
3. certification of information sources by the respondent that is forwarded by the institution
4. certification of data by the institution,
5. imposition of supervision on the respondent by the institution,
6. submission of a correction of published articles by the respondent, and
7. submission of a retraction of published articles by the respondent.

In addition, the NIH may take further administrative action, including:

1. modification of the terms of an award such as imposing special conditions, or withdrawing approval of the PI or other key personnel,
2. suspension or termination of an award,
3. recovery of funds, and
4. resolution of suspended awards.

The institution may impose additional penalties such as loss of employment, reassignment of personnel, or admission to a mentorship program for retraining" (1). If a penalty of some sort is levied, it should be proportional to the crime and follow due process of the law and regulations.

If someone is suspected or accused of misconduct, and for whatever reason, cannot convincing prove that the charge is wrong, then they should seek immediate help from colleagues, mentors, the institution or lawyers. Such charges can be very damaging.

Is there an appeals process? Yes. The process for contesting a decision is outlined on the ORI Web site. (From Ref 1)

CAUSES OF MISCONDUCT

Examination of cases of misconduct by the ORI suggests that there are seven clusters of problems, and some of these overlap. In general, these include: stresses and pressures, the culture of the employee organization, employment insecurities, personality factors, and others. Additional sources also support these as causes but some also stress a failure of the organization to provide proper support and guidance in research (2). All researchers should be trained in research integrity (3). Training means becoming familiar with definitions, discussing actual cases, and providing proper supervision for sound scientific procedures to mitigate misconduct (4). Training and continuing education in misconduct should be part of the organizational mission and culture. The more senior people in the field have a responsibility to train the more junior investigators, and the latter have a responsibility to diligently and courageously perform according to sound ethical standards.

As noted, examples of misconduct cases are excellent training tools and should be part of a training program in avoiding misconduct (2- 6).

PROTECTING WHISTLEBLOWERS

Federal employees who are whistleblowers are protected by federal law, and depending on the case, there may be other protections. While history has shown that protection of whistleblowers is not always effective (7, 8), protections for whistleblowers are in place. If someone thinks misconduct has occurred, they are required to report it. Confidentiality is essential in this process.

Universities have similar policies and other institutions are likely to as well.

ARE YOU CONSIDERING REPORTING MISCONDUCT?

While real misconduct is rare, everyone has a duty to report misconduct as noted above. A problem is that in many cases misconduct is not always simple and clear. Also, it is a serious accusation with potential and real consequences for the accused, and sometimes for the accuser as well. A study has shown that emotional damage to the accused can linger even if the accused is vindicated (9, 10). A number of studies have shown that whistleblowers are not always protected (11, 12). Thus, In order to minimize unnecessary harm to everyone, whistleblowing should be taken very seriously and done with great care. Institutions are committed to protecting whistleblowers and protective procedures should be in place. Some possible guidelines follow.

1. What exactly is the misconduct? Have you really observed it? In many cases, a reasonable first rule is to separate, in the observer's mind, what they see from what they think it means. Is it really misconduct or are other interpretations possible? This rule will help focus on the facts. Remember that intention matters. If there was no intention to cheat, then misconduct is unlikely – perhaps it is sloppiness or perhaps something else. But also remember that excessive sloppiness can be misconduct and create problems of many kinds.
2. Be careful of language. It may be regretted if someone uses words such as "making up data, plagiarizing, misconduct, or falsifying" without solid proof or without a judgment of misconduct by an authority. If someone suspects

that another is doing something wrong, remember that a suspicion is not proof.
3. Some believe that problems should be solved at the lowest level, that is, without going up the line of authority unless it is necessary. If someone feels that it must be discussed with another person, then a reasonable approach might be to discuss the alleged misconduct with a supervisor who is above the accused, or with a person designated and trained to handle such allegations confidentially. Many institutions have an office or a phone number to call where an anonymous conversation can take place. The charge of misconduct can then be investigated confidentially. This helps to protect both the whistleblower and the not-yet-proven-guilty researcher.
4. If, after going over the requisite precautions, someone still wants to proceed with a report, then they should find out the proper procedures at the respective institution. It should be possible to discuss the problem confidentially and perhaps anonymously. Again, distinguish what you see from what you think it means. It may be obvious that misconduct has occurred but it may not be so obvious in some cases. It is obvious that the procedures for determining if misconduct occurred must follow due process of law.
5. Many institutions have an office where complaints can be brought and discussed confidentially. Find such a place even if at another institution. If a designated person experienced in such matters feels that the accusation is warranted, then an investigation will be considered.

6. If the official, preliminary examination suggests that misconduct has occurred, something similar to the following may happen.
   a. The lab in question is shut down and the data notebooks are taken.
   b. A committee, that includes investigators familiar with the research topic, then investigates. The investigation can take a month, a year, or more.
   c. If an individual is found guilty, and many are not, then the consequences can be severe, such as termination of employment.

Note that a finding of misconduct does not automatically mean that the person no longer has a career in science. It will mean that the transgressor must think about why it happened and decide whether he/she wants to continue in science given the causes of the misconduct. If the person wants to continue, and if it is possible, then he/she will most likely be given a time of close supervision with requirements for retraining and rehabilitation.

A major deterrent to misconduct is continuing training in the topic.

HOW CAN RESEARCHERS HELP AVOID PROBLEMS?

As mentioned, a major deterrent to misconduct is continuing training so that there is a cultural awareness of possible problems. The researchers themselves can actively raise awareness of possible pitfalls and therefore help everyone in many ways. An obvious good practice is for everyone to show respect for and promote ethical practices by mentioning them and discussing them with colleagues.

Getting training in research misconduct is not a directive to go searching for and accusing wrongdoers. It could mean that in some situations, but a more important focus is for the training to improve the individual practices of the trainees. It is intended to help everyone in their own careers.

Researchers are ambitious, energetic, and passionate in their work and some behaviors may approach an ethical boundary that should not be crossed. Most researchers by far do not cross the boundary as misconduct is quite rare. But fellow researchers and colleagues can be there to provide private and gentle reminders and support. A determination of misconduct will obviously affect the wrongdoer, but it will also possibly affect the coauthors and fellow lab members of that wrongdoer.

CASES

Case 1. Too much irresponsible talk over a misunderstanding.

Consider a case where a lab member "A" watches another lab member "V" enter data from an electronic counter into his notebook. A happens to notice that the entries are not the same as those on the counter. A thinks V is making up or falsifying data. He tells this to others in the lab. Soon V realizes that everyone is avoiding him, and his supervisor seems angry. Finally the supervisor confronts V with the accusation that he is making up data. V is stunned because he was doing no such thing. After going over the facts, V agrees that he was not entering the exact numbers from the counter because he was mentally subtracting out a background amount from the data, which had to be done ultimately. He also said that what he was doing was only a preliminary curiosity; he planned on entering the printout from the counter and show in detail all of the calculations in his notebook. It took some time for the rumor to dissipate as it had spread to other departments. The inappropriate comment of "making up data" was not justified and created problems for the lab member and even the supervisor. Certain words or phrases should be avoided until they are for sure ok. Many situations such as this could be avoided with thoughtful discretion. After suspecting V, what should A have done?

Case 2. Can't repeat previous work.
Suppose you are a new post doc entering a new laboratory. While you have your own project, your advisor asks you to complete a project that was left unfinished because someone left the lab for a nice opportunity and advancement. However, after many tries you are unable to repeat the previous work and you really can't follow the notebook of the previous experimenter. You mention this to your advisor, and he says that the previous experimenter was an outstanding scientist and that you need to keep trying and practicing your technique; in other words, you are the problem, not the earlier scientist.

What can you do to resolve the problems which include repeating the data and gaining the trust of the advisor? Should you start mentioning possible "misconduct" to others in the group?

Case 3. Test for plagiarism.
Suppose your research is entering a relatively new area, and you decide that the best way to begin is to write a review article on the new topic. You decide that you want to include your staff in preparing the review because they need to know the material as well. You meet and divide up the tasks and areas, and everyone goes to

work on their particular job. As the review is being assembled, you become concerned that some of the less experienced staff members may not be citing existing material appropriately. You find at least one example where quotation marks and citations are missing. You are properly concerned about appearances of plagiarism.

When the review is finally assembled, can you search for passages that are the same as or similar to some in the literature? How can you do it? Are computer programs that search for repetitions available to you? How can you find them? Which are the best ones? Are they prohibitively expensive? It seems reasonable that if something is demanded of us, then we should have the tools needed to accomplish that task.

Case 4. Misconduct is a word that should not be used carelessly.

Which of the following is misconduct, and why or why not?

1. You have been put on a medication that interferes with cognitive abilities. When calculating and processing your data, you make some errors that get published.
2. You are a coauthor on a paper that required statistical analysis that is beyond your knowledge of the topic. Later it was found that the statistician was guilty of falsifying some of your data and analyses.
3. You have some emotional problems that sometimes make you volatile and reckless. Your part of a published report is later shown to be full of errors. You remember being impulsive and very angry at someone when you processed the data for that report.
4. You are under pressure to produce preliminary data for a grant application. When doing the calculations, you do

not include some of the numbers that seem to stand apart as being too large. You reason that this is only preliminary and more extensive experiments will be done anyway if the grant is funded.

SUMMARY

Misconduct has been defined and training in the subject is necessary. There are many instructive historical cases that can be examined. Procedures are in place for determining if misconduct occurs, and what remedial action should be taken. If there is a judgment of misconduct, an appeal process is in place. If someone knows of misconduct, they must report it; there are mechanisms and guidelines for reporting, and there are protections for whistleblowers. A major deterrent to misconduct is continuing training in the topic at all levels. A conviction of misconduct can seriously taint and hurt not only the careers of the convicted but also careers of innocent coauthors and fellow lab members.

REFERENCES.

1. https://grants.nih.gov/grants/research_integrity/research_misconduct.htm, accessed on September 26, 2016.
2. Davis MS, Riske-Morris M, and Diaz SR. Causal factors implicated in Research Misconduct: Evidence from ORI Case Files. Sci Eng Ethics, 13: 395-414, 2007.
3. https://sites.google.com/site/marshallshuler/science-and-society/scientific-misconduct/root-causes-of-scientific-misconduct/, accessed on September 29, 2016.
4. Dubois JM, Anderson EE, Chibnall J, Carroll K, Gibb T, Ogbuka C, and Rubbelke T. Understanding Research Misconduct: A Comparative Analysis of 120 Cases of professional wrongdoing. Accountability in Research 20: 320-338, 2013.
5. Resnik DB. From Baltimore to Bell labs: Reflections on Two Decades of Debated about Scientific Misconduct. *Accountability in Research*. 10: 2003, 123-135.
6. Titus SL, Wells JA, Rhoades LJ. Repairing research integrity. Nature 453: June 2008, 980-982.
7. Bird, SJ and Hoffman-Kim D. Whistleblowing and the Scientific Community. A special issue of Sci Eng Ethics 4(1), 1998.
8. For example, Joy, AB. Whistleblower. Bay Tree Publishing, 2010.
9. Lubalin JS, Matheson JL. The fallout: what happens to whistleblowers and those accused but exonerated of scientific misconduct. Sci Eng ethic s 5: 1999, 229-250.
10. http://www.the-scientist.com/?articles.view/article-No/18472/title/scientists-exonerated-by-ori-report-lingering-wounds/, accessed on September 29, 2016.

11. Lubalin, Gunsalus, C.K. (1998) How to Blow the Whistle and Still Have a Career Afterwards, *Science and Engineering Ethics* **4**: 51–64.
12. Sieber JE. Why fallout from whistleblowing is hard to avoid. Commentary on "The fallout: what happens to whistleblowers and those accused but exonerated of scientific misconduct" (J.S. Lubalin and J.L. Matheson). Sci Eng Ethics. 1999 Apr;5(2):255-60.

9

Conflicts of Interest and Commitment

Conflicts of Interest (COI) can undermine the credibility of an institution or researcher and even lead to serious penalties. Accordingly, COIs must be declared and managed. The public trust in science must be maintained.

A conflict exists when someone can derive personal gain by carrying out official actions. The concern is that the possibility of gain might corrupt the person's actions or judgment.

If someone has a COI, it does not mean that they are intrinsically bad or in trouble. It makes sense for the most expert or most involved investigators in a certain topic to contribute their expertise and use their knowledge. The presence of a COI does not mean that the investigator absolutely cannot do or receive benefit from the work. If there is a COI, then the activity/situation may have to be monitored and managed. But, there may be situations where a COI precludes working on the project. Thus, all possible COIs are obviously best declared and evaluated before the work begins. Much useful and productive work has been done in situations where COIs are declared and managed.

COIs must be declared so they can be assessed. The assessment will help protect everyone involved. Training in COI recognition and management is sometimes required of researchers, and an annual certification may be needed as well. Reviews of activities by officially designated COI committees can be required. The COI committees make decisions, which vary from the absence of a COI up to a serious COI where the investigator cannot be involved. The investigator should always check with the institution or funding agency to know his/her obligations.

There are a variety of ways that COIs can be managed. These include third party involvement for review or oversight, delegation of the specific duties where the conflict lies, and declaration of the COI. The details of the management plan are worked out, if possible, with the appropriate officials at the institution.

The NIH has important COI policies (1-4) and other research institutions usually have COI polices as well. If anyone is in doubt about an activity that may be a COI or appear to be a COI, it is best to get a decision from the appropriate office at their institution or from their funding sources as soon as possible.

FINANCIAL CONFLICTS OF INTEREST

One kind of COI, a financial conflict, is found when a person, or a family member, or organization, has multiple financial interests. It occurs when the judgment of an investigator could be biased in some study because of his/her financial interests. A COI also occurs when an immediate family member of a faculty member does business with the university; this could result in bias or favoritism for that business. COIs can involve various activities such as consulting, accepting various honoraria, being involved in startup companies, and more.

Sponsors of research expect objectivity and unimpaired judgment in their investigators and would be very concerned about unannounced COIs. The simple appearance of a COI may undermine the integrity of the involved person or even the organization where he/she is employed (1-3). Financial COIs must be clearly stated in grants, published papers, and in other presentations and documents as well.

The dangers of COIs have caught the attention of legislators. The Sunshine Act of 2010 requires that all makers and sellers of drugs, medical devices, biologic supplies and related materials follow and report all financial relationships with physicians. The report is made to the Centers for Medicare and Medicaid Services (CMS). The goal is to make such financial relationships known so that any COIs can be identified and dealt with (5, 6). Even a gift of small value to researchers may have to be reported.

COIS IN REVIEWING GRANTS AND CONTRACTS

An NIH grant application cannot be reviewed by someone who is in conflict unless it is excused. These include those who are named in the proposal or a member of an NIH advisory Council, or someone (or a family member) who would receive a financial benefit from the grant, or someone from the same institution. There are several situations where an investigator must leave the discussion or the room if they have a COI with a proposal undergoing review (7). All possible conflicts are best discussed with the grants mangers before the review. The rules are similar for contract reviews (8). Agreement to review any document implies that the reviewer can provide a fair and impartial review.

COIS IN AUTHORSHIP

Authors must state significant COIs in publications. The ICMJE has a conflict of interest (COI) form (9) that makes authors aware of and helps them identify any possible COI or an appearance of a COI. It asks if the authors or their relatives or their institutions have any financial interest in the published work through consulting agreements or patents or anything else.

COIS IN REVIEWING PAPERS SUBMITTED FOR PUBLICATION

Reviewers in conflicted situations should not review submitted papers unless the conflict is discussed with the editors and approved before the review. COIs in peer review have been discussed in Chapter 7. Financial COIs must be declared and other COIs must be as well. Other conflicts include reviewing work from others at the same institution, or work from recent collaborators. In some cases reviewers may have to recuse themselves from the review.

CONFLICTS OF COMMITMENT

An investigator working at a research institution has a commitment to that institution, and this can be compromised. For example, it is not appropriate to draw salary from the institution and then also make financial gains from another, unless it is approved by the institution. A conflict of commitment occurs when improper amounts of time and energy are directed away from a person's main job such as university teaching, and directed towards activities that offer additional pay such as consulting agreements (4). There are a variety of outside or extra activities that enhance the faculty member and his/her institution, such as serving on grant review committees. But these activities should not compromise or inhibit that faculty member's duties at his or her main employer. To avoid a conflict of commitment, an outside activity approved by the university should not require more than a certain maximum amount of time, perhaps one day per week.

To avoid conflicts of commitment, it is safest to report all activities apart from normal duties to the appropriate institutional official. Chairpersons or institutional policies can be guides and help inform decisions.

CASES

Case 1. Working out a conflict of commitment.
A faculty member is an expert in an area where litigation sometimes occurs. Most recently, she has been hired as a forensic expert on an important case. This commitment will require 3 full days in a specific week.

Does this violate the conflict of commitment rule of one day per week? How can this be considered and how can the faculty member proceed?

Case 2. Permissions for COIs.
A researcher is expert in vaccines and has an impactful laboratory that is involved in vaccine development. A company that also produces vaccines wants to hire the researcher as a consultant. Does the researcher have to report this consulting activity and get permission to do it? What are the usual time constraints in this activity? Is there any difference if the company pays money to the researcher personally, or if the money is given to the researcher for his lab research only?

Suppose the company gives the researcher funds to study one of their vaccines, and they later ask the researcher to lecture on the positive qualities of their vaccine. Is this a COI? How should it be handled?

Case 3. Manipulating a conflict of commitment.
A faculty member has taken a lucrative consulting agreement with a company. He realizes that it is going to take up a lot of time, so he assigns much of the work to various graduate students and postdocs so that his personal time is less than one day per week. Is this allowed? What should be done to be safe?

Case 4. When does an activity become a COI?
In the past, a senior faculty member has been given permission to have a significant financial interest in a startup company focused on developing new antidepressants. He/she is currently writing an NIH grant application whose goal is to develop new animal models of depression.

Could there be a COI there? How would that be determined? Does he have to declare his company involvement in the grant?

SUMMARY
Conflicts of interest or commitment occur when someone's fair judgment, impartiality, objectivity and time commitments might be compromised. Committees and rules to identify and deal with COIs have been established in government and at other institutions. All possible conflicts must be reported before the work begins and often permission must be given to proceed. COIs must be stated in grant applications, publications and other presentations. Training in COI recognition and management would be helpful.

REFERENCES.

1. https://ethics.od.nih.gov/Topics/coi.htm, accessed on March 28, 2017.
2. https://grants.nih.gov/grants/policy/coi/nih_review.htm, accessed on March 28, 2017.
3. https://ori.hhs.gov/education/products/rcradmin/topics/coi/tutorial_4.shtml, accessed on March 28, 2017.
4. https://grants.nih.gov/grants/peer/NIH_Conflict_of_Interest_Rules.pdf, accessed on April 1, 2017.
5. https://www.ama-assn.org/practice-management/physician-financial-transparency-reports-sunshine-act, accessed on May 4, 2017.
6. https://www.cms.gov/OpenPayments/Program-Participants/Physicians-and-Teaching-Hospitals/Registration.html?utm_medium=cpc&utm_source=Google&utm_campaign=Open_Payments&utm_term=sunshine%20act, accessed on May 4, 2017.
7. https://grants.nih.gov/grants/peer/Grant-Reviews-508.pdf, accessed on Aril 1, 2017.
8. https://grants.nih.gov/grants/peer/Contract-Reviews-508.pdf, accessed on April 1, 2017.
9. http://www.icmje.org/about-icmje/faqs/conflict-of-interest-disclosure-forms/, accessed on March 27, 2017.

10

Ethical Behavior Among Colleagues

This chapter is about interacting with colleagues in a fair and equitable way. Getting ahead in life often depends on personalities and interpersonal behavior as well as on scientific skills and technical knowledge. Collegiality can be important and is sometimes a requirement for employment (1).

Researchers who are unethical when dealing with colleagues are in danger of damaging their reputations and careers. While many say that discussing this topic of co-worker ethics is essential, there are few formal opportunities to do so. While this chapter is not a comprehensive review of the topic, it will address a number of relevant issues such as: What are the ethical topics or issues that arise among colleagues? What are the guidelines? How can collegial skills be improved?

TIME AND EFFORT FOR GROWTH

Developing or enhancing collegial skills can take time and effort. Therefore, a reasonable starting guideline is to devote some time for thinking about ethical issues among colleagues. Time and practice are needed to develop some personal rules and habits for co-worker ethics.

GIVE CREDIT WHERE IT IS DUE

Research is often done in a competitive atmosphere. Nothing stings more than when someone takes or is given the credit for what you have done. Of course this works both ways; so a rule would be to give credit to others, even for the little things that some might not notice. Or, if someone is credited with something and the credit should be shared with someone else, it is best done quickly. Supervisors need to share credit and recognize those who are supervised, and vice versa.

BEING A GOOD COLLABORATOR

At one point or another, many investigators enter into collaborations. They are useful because they expand the capabilities of individuals; teams are often more effective than individuals. But it may not be easy being a collaborator, and some traits are likely to make collaborations easier.

1. Being open to other's ideas – Sometimes researchers are very fixed in their vision, but someone else's ideas can be better or worth trying.
2. Being communicative – As part of a team, one is expected to contribute to the thinking and processing.
3. Being generous with others – Being generous to colleagues sets up a good working atmosphere; this doesn't mean that everyone won't get their fair share in the end.
4. Having a willingness to compromise – This is part of working in a team, part of making complementary skills work.
5. Bringing some needed skills to the table and maintaining them – Something has to be offered to be part of the team.

6. Having a comfortable working style – This makes working together easier.
7. Willingness to clearly commit – It seems best to have written agreements before collaborations actually begin regarding: data management, assignments, timetables, and positions in the author list of potential papers.

No one absolutely *has* to be a collaborator, and a person's style may not be suited to it, but collaborations can be powerful, especially in today's world where there are so many different skills and techniques that could be blended. Indeed, a collaborative style may be required by some employers. If somebody likes to work alone, if they don't like telling their plans to others, if they are more introverted than extroverted, then maybe they would be happier not being part of a team. Plenty of individuals make very nice contributions singlehandedly. But collaborations can be powerful and get things done that individuals could not.

DIVERSITY

Science is a multicultural endeavor. It is common for people with various differences to work together. They may be scientists of different races, nationalities, languages, ages, and sex. It is the duty of each person to be aware of their own personal jealousies and prejudices towards race, age, sex, and more, and to resolve them.

While individuals do differ in *performance*, which is evident in activities like the Olympics, they are *equal* under the law and ethical rules. Colleagues at work will be judged on the basis of performance, but they must be treated fairly. It is unethical to harm someone or to shortchange them on the basis of simple emotional dislike or aversion alone.

USE POSITIVE AND SUPPORTIVE LANGUAGE

Colleagues often make comments about other colleagues, but they are not always supportive or benign. If there is no reason to do otherwise, keep language supportive, or at least neutral or benign. Some people are automatically careful in how they talk about others, and they are a pleasure to be with. But some are not. The two different conversations, one supportive and the other not, can sound very different and leave listeners with a very different feeling about themselves and another person (2). Consider the example of the two conversations at the end of this chapter (Case 2).

Detraction is dragging up unattractive things that happened in someone's past. Unless the past is having an active, negative impact on the present, it is probably best left alone. One could argue that it is not fair; everyone deserves a second chance. It's best to be careful of such gossip about someone's past.

PRACTICE THE GOLDEN AND PLATINUM RULES

Treat others as they would like to be treated. Everyone would certainly like this. Also, if one's actions impacts on others, the rule should be "First do no harm or do as little harm as possible." (1)

Because people often have an automatic or subconscious response to another person, a good rule is that simple dislike of another can never justify destructive behavior towards that person. Again, a good practice is to treat others as they want to be treated.

CONFLICTS AMONG COLLEAGUES

Conflicts happen. While we may not like them, they are normal and often needed to make progress. In many cases, it is possible to limit the negative outcomes of conflict and to enhance the positive or productive outcomes. As noted above, seeking out training in non-violent communication and conflict resolution can help (3, 4). Training in other skills such as negotiating can be helpful as well (5, 6). Also, training in relational skills can help; do we think of others as objects to be used, or as respected human beings (7, 8).

Perhaps a special mention is due when conflicts involve an imbalance of power, such as a conflict between a student and a faculty member, or an employee and supervisor. Conflicts with an imbalance of power can be difficult. Sometimes when things aren't going well, employees blame the supervisors, but, we need to respect supervisors because their job can be difficult. They have the responsibility for productivity. We sometimes hear the complaint against supervisors that "It needed to be done but I didn't like the way it was done." But we need to be tolerant as supervisors are only people with differences in preferences and style. Supervisors may have more information than others and need leeway and understanding in their actions. On the other hand, having supervisory power requires responsible actions, and all supervisors should respect their power and impact on others. Fairness, responsibility, and sometimes selflessness often go along with the higher status of being a supervisor.

AVOID BRUISING COLLEAGUES

Sometimes, the personality traits that help us be good scientists, such as aggressiveness and passion, can create some problems with colleagues. Being appropriately aggressive with research can result in the side effect of bruising or insulting colleagues. Some researchers may seem brusque, thoughtless and avoidant of others. It seems that a good rule is to be careful of injuring someone, and to be quick with apologies and amends. Also, because science is very multicultural and international, care should be taken to avoid cultural slights or problems. The human side of collaborating needs attention and is unavoidable (7, 8).

HUMAN NATURE: BOTH POSITIVE AND NEGATIVE INFLUENCES

Researchers need to be aware of some of the many, less rational and more automatic tendencies that humans possess (9). They are part of our basic human nature which has developed over a complex evolutionary path. Dealing with and trying to understand our complex human nature can humble the very brightest. Many human tendencies and traits have evolved to deal with ancestral threats, and therefore exist for good reasons overall. But they can create problems in the present day world. A message here is that they are only *tendencies and emotions* based in our inherited nature. They are *not* rules or commandments for behavior that must be followed today. People can respond to situations in many different ways. While understanding these may have nothing to do with one's research, insight into human nature will influence everyone's interactions with others. Reasonable goals are to be positive, supportive and productive. The following paragraphs list some of these tendencies.

It is often assumed that all researchers are eminently reasonable and rational, but in fact, like other people, they are not so all the time. They may have a confirmation bias which means that if they have a preconceived notion or belief, then they will tend to seek out and agree with the data that supports their belief (10, 11). They will ignore the data that does not. Suppose someone believes that "All politicians are crooks." They will say "I told you so." every time there is a news article that describes corruption. But they may totally ignore all of the articles that point out how selfless and committed many public servants are. In a profession where opinions should be based on facts and sound judgment, this kind of bias can be a problem.

Consider what are called attribution errors (12). If someone sees something happen to another person, perhaps they simply trip over a carpet; the observer tends to think of that person as responsible for it and feels that they personally are clumsy or accident prone. But if the observer does it, he/she thinks of the trip-up as a problem with the placement or nature of the carpet rather than their own actions. The point is that people erroneously tend to attribute accidents or problems to the deficient character or personality of the victim, whereas in reality other circumstances may be the fault. This tendency can be a factor in how colleagues react to others, or in how problems are acknowledged and handled, or in how responsibility for events is assigned. A more thoughtful position is to seriously consider the circumstances rather than the character or personality of the person involved.

Suppose two colleagues, A and B, have different views about something. But if one of the colleagues, say colleague A, for whatever reason, says that he/she agrees with the other colleague B, then A moves closer to agreeing with B's viewpoint. The simple act of saying you agree with someone else (even if you may not) may result in moving closer subconsciously to the belief of the other. Beliefs can erode by speaking something contrary to one's beliefs (13, 14). In a system where one's own ideas can be the currency for getting ahead, one may need to be careful of this.

Suppose there is an argument and the focus is on being either right or wrong. Then it seems that someone will have to lose, i.e. be wrong. Can the picture be shifted by arguing over two different points of view such that both are right from different perspectives? In that case, the outcomes may be the best focus of the argument.

Many additional natural tendencies and traits that affect interactions with others could be described. It is worth paying attention to this topic going forward, and to study it further. Kahneman is one who has explored this (9). Our human nature affects us, sometimes thwarts us, as we strive to be ethical, fair, and wise with our colleagues. Interesting topics for further study include the need to get ahead, schadenfreude, authority, biases and more biases, serving our fixed and sometimes irrational ideas, sociopathic colleagues, developing courage, and more.

More can be said about courage. Do we have the courage to choose ethical actions? Do we go along and get along, or do we stand up for ethically correct alternatives? Developing courage has been discussed (2, 15)?

TAKE SOME TIME BEFORE RESPONDING.

Responding emotionally or flippantly without thinking can sometimes create problems. Maybe it is best to go slowly, and think about it.

Research has revealed that our mental processes can be thought of as composed of two systems: one of these is fast in that it reacts very quickly or automatically, and the other is slow. The fast system almost automatically provides answers to some questions very quickly, with the emphasis on quickly. The question might be how much is 1 and 1. Or an action might be grabbing a potato chip in response to slight hunger. It is almost like a reflex. It has been speculated that this fast responding system has been important for surviving sudden threats in our ancestral past, and that seems reasonable. But there is also a slower system. The slower system is more rule-and thinking- based and produces planned or thought-out behavior. It might try to calculate the odds of surviving some complex situation. It can only help if one becomes aware of this, and gives themselves *time* to respond in the best way (16-18). Give the slower and more thoughtful side a chance to work. The well-known aphorism of "Let's sleep on it" acknowledges the known advantages of taking time to reply.

UNETHICAL COLLEAGUES

What can be done about unethical colleagues? What if someone's very ethical behavior is being used by others to gain an unfair advantage? Perhaps they won't help others in the lab even though they are getting much help from colleagues. Perhaps they find ways to avoid giving deserved credit to others. Depending on the situation, one may be able to address it with the other person or with a supervisor. But if it falls in the realm of the serious ethical problems like those covered in the earlier chapters, then there are protocols in place for handling various kinds of misconduct.

Personality tendencies that are sometimes problematic are found in everyone, but some kind of protection may be needed from others who are consistently biased or unfair (Chapter 7 in ref

15). Seeking advice on how to deal with problem colleagues may be needed from time to time.

Who should promote or enforce ethical behavior among colleagues? Perhaps we all should. We can gently notify those who cross ethical boundaries.

CAN WE LEARN TO BE BETTER COLLEAGUES?

Some say that "People don't change," or "That's just the way I am." But psychologists have obviously proven that behavior can be modified. New skills and responses can be learned. A somewhat extreme example is how policemen and firemen are trained to overcome their fear while carefully carrying out their mission. Time and practice spent on learning to be a good colleague and coworker is likely to be helpful. Set some goals for being a good colleague, discuss them with others, discuss examples, read on the topic, and seek mentors when needed. Part of this learning may be an awareness of how often you tend to use people as objects rather than treat them as genuine people (7, 8, 15, 18). Continued training is helpful and the courage to gently deal with problem colleagues is also helpful.

CASES

Case 1. Dealing with a difficult mentor.
You have recently joined a new laboratory for post-doctoral training. You are excited about the project but your opinions about how to proceed seem to conflict with those of your mentor. More and more, it seems like your mentor is disapproving of almost everything that you do. You are puzzled, uncomfortable and a little resentful, but you don't want to leave the lab.

Research Ethics in the Life Sciences

What are the possible courses of action of the trainee? In terms of the imbalance of power, what are the realities of the situation?

Case 2. It matters how you say it and the words you use.

"A colleague has come under investigation regarding expenditures of his grant funds. He has a flawless track record and this investigation has surprised everyone. A group of faculty in the dining hall are discussing this (the colleague under investigation is not present).

"Hey Fred, you know this guy best, what did he do?"

Fred replies, "I have no idea."

Someone else says "He must have done something to get investigated."

"Yeah, didn't he get a new car recently?" and everyone laughs at the intended humor.

"He hasn't shown his face lately."

"Yeah, sounds like guilty behavior to me."

"I'm going to stay away from him or else I might get investigated."

There are a few serious nods and introspective faces around the table.

Now consider this alternative conversation.

"Hey Fred, you know this guy the best, what is happening?"

"I'm not sure. The alleged charges are being investigated, so I don't know. Maybe he did something bad. But, it could be a misunderstanding or maybe something minor. I've seen that before."

"Where's he been? We haven't seen him."

"I think he feels bad and maybe he's ashamed because of the investigation. But I think I'll give him a call and invite him to join us one of these days. Not so we can grill him, but so he can explain if he wants to. If he has made a mistake, then I'd like to learn from it.""

It is obvious which conversation is more supportive. Does competition in the workplace – a positive force in many ways – set us up to be harsh to others? Probably so. Does our human nature have a tendency to deflate others in a self-aggrandizing way, and even to enjoy their misfortune (schadenfreude)? Perhaps yes. But again, our natural tendencies may not always be appropriate. (Case taken from Ref 2)

Case3. Fairness in reporting to the press.
You are a member of a relatively large collaboration team that has successfully published some great work. For whatever reason, the Institute leadership has asked you to meet with the press at a formal press conference to explain the work. You are inclined to do this because sharing with the public is a good thing particularly if public funds support the research. In preparation for the conference, you see that none of the collaborators are mentioned by name in the press release and in the conference materials, although there is a reference to the published materials. However, your name is clearly and prominently given.

Is this acceptable? The PR office says that including all the names is not practical, and the press wants only information without frills. What are your options? What are the minimum requirements for you to go ahead with the conference? Or, do you just accept their wishes, and proceed the easiest way? How important is giving credit?

Case 4. Evaluating colleagues.
Your career has been successful, and you are in a position where others ask for your opinion of various colleagues. You are asked "What do you think of them?"

What are the main criteria that you use to judge a coworker or employee? Do they include grades in graduate school? Do they include the results of an IQ test in college? Some feel there is only one major issue; what is it? When should you discuss collegial skills?

Case 5. A program for collegial behavior.
Because there have been many interpersonal problems at your institution that involves ethical behaviors, you have been asked to help set up a professional development program that will in part focus on co-worker ethics.

How would you teach collegial/co-worker ethics? Would you use cases? What about asking for cases from individuals who have been affected by the problems? What else? What would you include in the curriculum? What existing tools would you use?

SUMMARY

While researchers are all different individuals, they all possess various common human traits that is their human nature. Because these traits impact on co-worker interactions, it is useful to make an effort to understand them and to select behaviors that facilitate rather than inhibit constructive co-worker interactions. Some behaviors that should help with colleagues are to give credit where it is due, to try to be a good collaborator, and to exhibit a variety of additional collegial skills. One also needs to be aware of and careful of some human traits that include various biases and errors. Being destructive to another simply and solely because one doesn't like that person is not acceptable. When confronted with an issue, taking enough time to evaluate and to respond effectively is recommended.

REFERENCES

1. Freedman S. Collegiality Matters: How do we work with others? (2009) Proceedings of the Charleston Library Conference. http://docs.lib.purdue.edu/cgi/viewcontent.cgi?article=1054&context=charleston, accessed on September 24, 2016.
2. Kuhar MJ. Collegial Ethics: What, why, and how. Drug Alc Depend 119 (2011) 235-238.
3. CNVC, 2011. http://www.cnvc.org, accessed on September 19, 2016.
4. CR: Conflict Resolution Skills, http://ctb.ku.edu/en/table-of-contents/implement/provide-information-enhance-skills/conflict-resolution/main, accessed on September 19, 2016.
5. What is Win–Win Negotiation? (2010). Accessed at http://www.negotiations.com/articles/win–winsettlements/on September 19, 2016.
6. Fisher R, Ury WL, Patton B. *Getting to yes: Negotiating Agreement without Giving In.* Penguin Books. New York, NY 2011.
7. https://arbinger.com/about/, accessed on April 8, 2017.
8. The Arbinger Institute, *Leadership and Self Deception*, Berrett-Koehler Publishers, Inc. San Francisco CA. 2010.
9. Kahneman D, *Thinking fast and Slow*. New York. Farrar, Straus and Giroux. 2011.
10. Lodge M, Strickland AA, Taber CS. Motivated reasoning and public opinion. *J Health Politics, Pol law.* 36 (Dec 2011): 935-944.
11. Mercier H, and and Sperber D. Why Do Humans Reason? *Behav and Brain Sci.* 34: (April 2011) 57-74, 74-111.

12. http://study.com/academy/lesson/fundamental-attribution-error-definition-lesson-quiz.html, accessed on September 24, 2016.
13. Carlsmith JM, Festinger L. Cognitive consequences of forced compliance, J Abnorm Soc Psychol, 58 (1959) 203 – 210.
14. Gawronski B and Strack F. On the Propositional nature of Cognitive Consistency. J Exp Soc Psychol 40 (2004) 535-542.
15. Kuhar, MJ. *The Art and Ethics of Being a Good Colleague.* CreateSpace Independent Publishing Platform, North Charleston SC. 2013
16. Two System Thinking. http://timreidpartnership.com/Site/An_introduction_to_2_system_thinking.html, accessed on September 19, 2016.
17. Of Two Minds When making a Decision. http://www.scientificamerican.com/article/of-two-minds-when-making/, accessed on September 19, 2016.

http://www.collegialethics.com, accessed on April 4, 2017.

Notes

Notes

Notes

About the Author

Michael J Kuhar, PhD., is at Emory University in Atlanta, where he is a Senior Faculty Fellow in the Center for Ethics, a Candler Professor of Neuropharmacology at the Yerkes National Primate Research Center and School of Medicine, and a Georgia Research Alliance Eminent Scholar. He is a researcher who has worked and published widely for more than forty years (http://www.michaeljkuhar.com/). His work has been recognized by a number of awards. His numerous trainees are found in a variety of positions in government, academia, and the drug industry around the world. He is the author of *The Art and Ethics of Being a Good Colleague* and has published and lectured on a variety of topics in research ethics.